Ina Klassen

Charged colloids as model systems for metals

Ina Klassen

Charged colloids as model systems for metals

Short-range order phenomena and nucleation behavior

Südwestdeutscher Verlag für Hochschulschriften

Impressum/Imprint (nur für Deutschland/ only for Germany)
Bibliografische Information der Deutschen Nationalbibliothek: Die Deutsche Nationalbibliothek verzeichnet diese Publikation in der Deutschen Nationalbibliografie; detaillierte bibliografische Daten sind im Internet über http://dnb.d-nb.de abrufbar.

Alle in diesem Buch genannten Marken und Produktnamen unterliegen warenzeichen-, marken- oder patentrechtlichem Schutz bzw. sind Warenzeichen oder eingetragene Warenzeichen der jeweiligen Inhaber. Die Wiedergabe von Marken, Produktnamen, Gebrauchsnamen, Handelsnamen, Warenbezeichnungen u.s.w. in diesem Werk berechtigt auch ohne besondere Kennzeichnung nicht zu der Annahme, dass solche Namen im Sinne der Warenzeichen- und Markenschutzgesetzgebung als frei zu betrachten wären und daher von jedermann benutzt werden dürften.

Verlag: Südwestdeutscher Verlag für Hochschulschriften Aktiengesellschaft & Co. KG
Dudweiler Landstr. 99, 66123 Saarbrücken, Deutschland
Telefon +49 681 37 20 271-1, Telefax +49 681 37 20 271-0
Email: info@svh-verlag.de
Zugl.: Bochum, U, Diss., 2009

Herstellung in Deutschland:
Schaltungsdienst Lange o.H.G., Berlin
Books on Demand GmbH, Norderstedt
Reha GmbH, Saarbrücken
Amazon Distribution GmbH, Leipzig
ISBN: 978-3-8381-1553-5

Imprint (only for USA, GB)
Bibliographic information published by the Deutsche Nationalbibliothek: The Deutsche Nationalbibliothek lists this publication in the Deutsche Nationalbibliografie; detailed bibliographic data are available in the Internet at http://dnb.d-nb.de.

Any brand names and product names mentioned in this book are subject to trademark, brand or patent protection and are trademarks or registered trademarks of their respective holders. The use of brand names, product names, common names, trade names, product descriptions etc. even without a particular marking in this works is in no way to be construed to mean that such names may be regarded as unrestricted in respect of trademark and brand protection legislation and could thus be used by anyone.

Publisher: Südwestdeutscher Verlag für Hochschulschriften Aktiengesellschaft & Co. KG
Dudweiler Landstr. 99, 66123 Saarbrücken, Germany
Phone +49 681 37 20 271-1, Fax +49 681 37 20 271-0
Email: info@svh-verlag.de

Printed in the U.S.A.
Printed in the U.K. by (see last page)
ISBN: 978-3-8381-1553-5

Copyright © 2010 by the author and Südwestdeutscher Verlag für Hochschulschriften Aktiengesellschaft & Co. KG and licensors
All rights reserved. Saarbrücken 2010

Contents

1 Introduction — 3

2 Physics of colloidal dispersions — 5
- 2.1 Soft condensed matter: colloids — 5
- 2.2 Model systems for atomic systems — 5
- 2.3 Interactions in colloidal systems — 7
  - 2.3.1 Attractive interactions — 7
  - 2.3.2 Repulsive interactions — 7
  - 2.3.3 The Poisson-Boltzmann equation — 8
- 2.4 Phase behavior of charged colloids — 11
- 2.5 Scattering theory — 13
  - 2.5.1 Light scattering — 13
  - 2.5.2 X-ray diffraction and light scattering — 18
  - 2.5.3 Model for short-range order — 19
  - 2.5.4 Local order in undercooled liquids: metals *vs.* colloids — 19
- 2.6 Kinetics of crystallization — 21
  - 2.6.1 Classical nucleation theory — 21
  - 2.6.2 Kinetics of homogeneous nucleation — 23
  - 2.6.3 Transient nucleation — 24
  - 2.6.4 Diffusion — 25
  - 2.6.5 Crystal growth — 26
- 2.7 Theory of elasticity — 28
  - 2.7.1 Elasticity of crystalline systems — 28
  - 2.7.2 Elasticity of colloidal systems — 30
  - 2.7.3 Shear waves in colloidal solids — 32

3 Experimental techniques — 35
- 3.1 Interaction control in charge stabilized colloids — 35
- 3.2 Multi purpose light scattering instrument — 37
  - 3.2.1 Static light scattering — 39
  - 3.2.2 Dynamic light scattering — 41
  - 3.2.3 Torsional resonance spectroscopy — 42
- 3.3 Time-resolved microscopy — 44
- 3.4 Ultra Small Angle X-ray Scattering — 46
  - 3.4.1 Beamline BW4 at HASYLAB (DESY) — 46
  - 3.4.2 Evaluation of diffraction measurements — 49

4 Results and discussion — 53
- 4.1 Interaction control in charged sphere silica systems — 53
- 4.2 Phase behavior and characterization — 55
- 4.3 Short-range order in charged sphere colloids — 63
  - 4.3.1 Short-range order of charged colloidal melts at different interaction strengths — 64
  - 4.3.2 Determination of the short-range order in the meta-stable state — 64

		4.3.3	Short-range order of charged colloidal melts near the solid-fluid phase boundary	67
		4.3.4	Short-range order of charged colloidal melts in comparison to metals	70
	4.4	Nucleation of colloidal crystals		74
		4.4.1	Crystallization kinetics	78
		4.4.2	Analysis of the nucleation kinetics within the framework of classical nucleation theory	81
		4.4.3	Transient nucleation	84
		4.4.4	Comparison to metals	89

5 Summary and Outlook   **95**

Bibliography   **99**

Publications within the thesis   **109**

Danksagung   **111**

Chapter 1
Introduction

Metals are non-transparent systems consisting of atoms, which show fast relaxation dynamics in the liquid state. Therefore, it is extremely difficult to directly observe e.g. solid-liquid interfaces, nucleation of crystalline phases and instabilities of a solidification front. In contrast, colloidal suspensions are transparent and the dynamics of their particles is much more sluggish than the relaxation of atoms in metals. Compared with atomic systems the elementary particles of colloidal systems are by several orders of magnitude larger than atoms. Colloidal systems consist of mesoscopic particles (1nm-10μm) dispersed in a carrier medium whose atoms or molecules are by orders of magnitude smaller than the colloidal particles. Colloidal particles are small enough to display Brownian motion, which prevents sedimentation under gravity, and large enough to allow for describing the solvent as a continuous and homogeneous background. Typical length scales (e.g particle diameters, lattice constants) are found to cover the wavelengths of visible light. Since colloidal systems are optically transparent, phenomena can be investigated by optical diagnostic means, which are not applicable in case of non-transparent metals on atomic length scales. Powerful and novel light scattering techniques including static and dynamic scattering methods in addition to torsional resonance spectroscopy were developed for characterization of new colloidal materials.

However, optical methods are in general limited due to turbidity and multiple scattering effects at high particle number densities. In addition, with increasing particle concentrations also the interparticle distances become smaller. To avoid multiple scattering and to cover an appropriate range of momentum transfer, radiation sources of smaller wavelengths have to be used. Sources of wavelengths in the ultraviolet regime do not come into question due to their strong absorption in colloidal systems. Therefore, X-ray scattering adapted for studies of colloidal systems is an appropriate scattering technique for turbid systems and small lattice spacings. To access larger particle concentrations without multiple scattering effects and also to obtain a larger range of scattering vectors that allows to determine the melt structure optical measurements are complemented by time-resolved X-ray Scattering using high intensity synchrotron radiation. The enlargement of the accessible range of momentum transfer q by using X-rays implies scattering at small angles. The corresponding Ultra Small Angle X-ray Scattering (USAXS) experiments were performed at the soft matter beamline BW4 at HASYLAB (DESY) in Hamburg.

In particular, charged stabilized colloidal suspensions consisting of colloidal spheres (macro-ions) suspended in solutions of small ions (micro-ions) were used for systematic investigations. Their interaction is described by a potential that contains only a repulsive term opposite to, e.g. the Lennard-Jones potential. But on both systems, colloids and metals it is assumed that the interaction potential acts isotropically. The interaction of charged colloids is evaluated by the linearized Poisson-Boltzmann theory for point charges. The finite size of the particles in a colloidal suspension is taken into account by a renormalization of the total surface charge Z_{bare} of the colloidal particles. The potential and therefore the range and strength of repulsion can be experimentally tailored via the particle's effective charge, the concentration of the screening electrolyte and the number density of particles. As an important prerequisite for systematic investigations on solidification a precise interaction control is guaranteed.

The scientific target of the present thesis is to analyze the potential of colloids as model systems with respect to the solidification of metals. For the complete understanding of the solidification process, the knowledge of the short-range order in the fluid state is of great significance because the short-range order influences both the nucleation behavior and crystal growth. Since the pioneering work by David Turnbull it is well known that liquid metals can be cooled far below their equilibrium melting temperature without their solidification. A liquid cooled below its melting temperature is denoted as an undercooled melt [1]. Due to small length scales in atomic systems, the microscopic process that suppresses the transition to the more favorable stable crystalline phase cannot be observed directly and a full understanding of local ordering in undercooled melts is still lacking. In 1952, Frank [2] postulated that the short-range order in undercooled metallic melts is consisting of icosahedral aggregates. Icosahedra, however, are characterized by a fivefold symmetry that is not compatible with the translational invariance of a crystalline material. Therefore, the icoshadral short-range order has to be broken before a melt can transform to a crystalline solid. This requires an energy and Frank has attributed this energy barrier to be the physical origin why metallic liquids can be undercooled below their melting temperature. Frank showed that the energy of an icosahedral aggregate of 13 atoms interacting by Lennard-Jones like potential is 8% lower than that of crystallographic clusters with the same number of atoms (e.g. aggregates of fcc or hcp structure). The energy calculations for an icosahedral aggregate include no explicit assumptions on the degree of undercooling or metastability. Frank's hypothesis does not exclude the preference of icosahedral type of short-range order in stable liquids above the melting point.

The present work focuses on a complementary approach to the issues involved with the solidification process by the use of colloidal model systems. Model systems are often used to describe equilibrium properties of simple fluids and solid materials. Recently however, interest has shifted from equilibrium situations to phenomena and processes occurring far from equilibrium. To proof its model character for metals the results of the short-range order in the metastable colloidal melt are compared in place of pure metals to those obtained for an undercooled Nickel melt [3] revealing an icosahedral short-range order in both systems, charge stabilized colloids and pure metals, respectively. In metals a fixed potential is characteristic for each element. Colloidal systems provide an opportunity of interaction tuning in a simple way leading to a continuous change from soft and long-range to a short-range hard-sphere like repulsion if an excess of screening electrolyte is added. The change of interaction also leads to a change of short-range order with fcc like structure in the state of hard sphere like behavior.

The formation of the short-range order in the undercooled state represents a pre-stage for nucleation. The extraction of interfacial energy as a function of undercooling parameters from nucleation experiments within the framework of classical nucleation theory is employed following a formalism with adaption to colloidal systems [4,5]. Taking account of transient nucleation effects, these results can be compared to metals by constructing a so-called Turnbull plot to discuss the model character of colloids concerning nucleation phenomena [6]. While the model character of charged colloids related to the short-range order is clearly evidenced to describe the behavior of pure metallic melts out of equilibrium state, the nucleation behavior described by the classical nucleation theory (CNT) should be discussed more critically. Discrepancies are observed between experiments and theoretical assumptions. Still many important open questions exist, ranging from details of the phase behavior to reactions involved in solidification. Many are parallel to those prevailing in the solidification of metals. Therefore, besides answering these questions for charged colloids an important task of this work is the continually critical assessment of the model character of colloidal spheres.

Chapter 2

Physics of colloidal dispersions

2.1 Soft condensed matter: colloids

Soft Matter is a branch of Condensed Matter Physics which deals with materials that display a strong response to weak external perturbations, i.e. they are characterized by weaker interactions of their elementary particles than hard materials. They facilitate the access to structural order phenomena on length scales that are much larger than those of a single molecular or atomic system. Soft matter comprises self-organizing supramolecular systems such as polymers, colloids, biopolymers, gels, liquid crystals, condensed molecular films, self-assembling amphiphilic systems and many other systems.

Colloidal systems consist of mesoscopic particles (1nm-10μm) dispersed in a carrier medium whose atoms or molecules are by orders of magnitude smaller than the colloidal particles. Colloidal particles are small enough to display Brownian motion, which prevents sedimentation under gravity, and large enough to allow for describing the solvent as a continuous and homogeneous background. Independently of the state of aggregation, a variety of colloidal systems is possible: colloidal emulsions (droplets in a liquid), suspensions (solid particles in a liquid), foams (gas in a liquid) etc. Typical examples of colloids are milk (emulsion) and gloss paint (suspension).

2.2 Model systems for atomic systems

Due to their mesoscopic length scales, colloidal suspensions are ideal model systems to address fundamental issues in condensed matter physics such as liquid ordering, crystallization and glass formation. The analogies and differences between the phase diagrams of colloidal suspensions and atomic systems and the corresponding structural and dynamic properties of the various systems as a function of the interaction potential have been thoroughly investigated [7]. A big advantage in using charged colloidal suspensions is that their interaction potential can be systematically tuned, opposite to metallic systems. Tunable interaction potentials play an increasingly important role for model systems to study a variety of phenomena in condensed matter physics. In the case of charged stabilized colloids, the repulsive range of the inter particle interaction potential can be controlled. The softness of the interaction is tuned in these colloidal suspensions by varying the solute or salt concentration. The typical colloidal length and time scales, i.e. microns and seconds, make it also possible to directly observe colloidal particles by optical means in real-space and real-time using video or confocal microscopy. Advanced colloid chemistry techniques are available to tune the chemical and physical properties of the particles or even to develop completely new and unique colloidal model systems. Colloidal systems are also interesting and relevant, as they find numerous applications in several industrial branches such as coatings, food, cosmetics but also in more technical applications as photonic crystals and data storage devices.

Characteristic for colloidal systems is the Brownian motion of their particles which arises from collisions between particles and solvent. As a result, the kinetic energy E_{kin} of colloidal particles is comparable to the kinetic energy of the solvent particles. That means that the kinetic energy of the

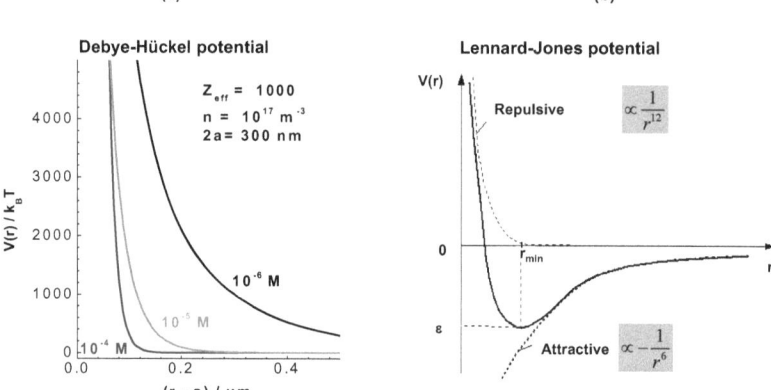

Figure 2.1: Interaction of charged colloids is successfully described by the Debye-Hückel potential (left), for the interaction of pure metals on the other hand a Lennard-Jones potential is frequently used (right). Variation of salt concentration in solution allows the tuning of the Debye-Hückel potential from a long-ranged soft (10^{-6} molar salt concentration) to a short-ranged hard sphere-like potential (10^{-4} molar salt concentration). Lennard-Jones potentials are fixed in dependence on the system. In equilibrium state, the pair of atoms or molecules tends to go towards a separation corresponding to the minimum of the Lennard-Jones potential ϵ at the minimum distance $r_{\min}$.

colloids follows the Boltzmann distribution,

$$p \propto \exp\left(-\frac{E_{\mathrm{kin}}}{k_{\mathrm{B}}T}\right) \tag{2.1}$$

with k_{B} the Boltzmann constant and T the temperature. The kinetic energy of atomic particles is distributed according to (2.1) as long as quantum fluctuations can be neglected. This assumption is valid if the temperature is sufficiently large and thermal fluctuations outweigh the quantum fluctuations. Due to the thermodynamic analogy between colloids and atomic systems, it is then possible to study phenomena in colloidal suspensions, which are not accessible in atomic systems.

The present thesis aims at scale bridging of structural investigations on colloidal systems with respect to metallic systems. Charged colloids interact via a pure repulsive potential due to the equal charge of the particles. This kind of interaction is successfully described by the Debye-Hückel potential and is discussed in sec. 2.3.3 in detail. Variation of salt concentration in solution allows the tuning of the Debye-Hückel potential from a long-range soft repulsive (10^{-6} molar salt concentration) to a short-range hard sphere-like potential (10^{-4} molar salt concentration) as demonstrated in Fig. 2.1 (a). The interaction of pure metals is frequently described by the Lennard-Jones potential, which in addition to a repulsive also has an attractive term. It reads as:

$$V_{\mathrm{LJ}} = \epsilon\left[\left(\frac{r_{\min}}{r}\right)^{12} - 2\left(\frac{r_{\min}}{r}\right)^{6}\right], \tag{2.2}$$

where $r_{\min}$ is the interatomic distance at which the potential becomes a minimum and ϵ is the specific Lennard-Jones parameter. When the separation r is very small, the $\frac{1}{r^{12}}$ term dominates, and the potential is strongly positive. Hence the $\frac{1}{r^{12}}$ term describes the short-range repulsive potential at small separations. In contrast the $\frac{1}{r^{6}}$ term becomes dominant at larger distances.

The Lennard-Jones potential is weakly attractive as two molecules or atoms approach one another from a distance, but strongly repulsive when they approach too close. The resulting potential is shown in Fig. 2.1 (b). At equilibrium state, pairs of atoms or molecules tends to go towards a separation corresponding to the minimum of the Lennard-Jones potential ϵ at the minimum distance $r_{\min}$.

2.3 Interactions in colloidal systems

Colloidal particles dispersed in a solvent undergo Brownian motion due to random collisions with solvent molecules. This counteracts gravitational forces which would lead to sedimentation. The particles have thermal (kinetic) energies in the order of $k_\mathrm{B}T$. During the process of this motion they may approach each other and attractive forces affect the particles. Under the influence of attractive forces, colloidal particles tend to aggregate to larger clusters [8]. Therefore other compensating repulsive interaction forces are necessary to stabilize the suspension. Two stabilization methods are commonly employed: charge and steric stabilization. Because this thesis is concerned with systems of charge stabilized colloids, only a brief introduction is given to steric stabilization.

2.3.1 Attractive interactions

Colloids in general interact via attractive van der Waals forces whose origin is due to fluctuating dipoles in a neutral material. The van der Waals interaction of two spherical particles of a radius a separated by a center-to-center distance r is given by

$$V_\mathrm{vdW}(r) = -\frac{A}{6}\left[\frac{2a^2}{r^2-4a^2} + \frac{2a^2}{r^2} + \ln\left(\frac{r^2-a^2}{r^2}\right)\right], \qquad (2.3)$$

where the Hamaker constant A considers material properties and depends on dielectric functions and polarizabilities of the particles and the solvent. The $V_\mathrm{vdW}(r)$ interaction scales with $-r^{-6}$ at large distances and with $-(r-2a)^{-1}$ as r becomes comparable with the diameter of the particles $2a$ [9–11].

If a suspension contains in addition to the colloidal particles also a smaller species (polymers or smaller particles), a second kind of attractive interaction occurs. It is known as the depletion interaction. The smaller particles exert osmotic pressure[1] on the bigger colloidal particles, forcing them to approach each other [12, 13].

2.3.2 Repulsive interactions

The two common mechanisms for repulsive stabilization in colloidal systems are steric and charge stabilization. Colloidal particles which are sterically stabilized are surrounded by a polymer layer as it is shown in Fig. 2.2. A considerable number of polymer chains is located at the surface either chemically bound or physically adsorbed. In the latter case, the bonding strength has to be larger than the thermal energy, otherwise the layer of polymer chains is not stable.

If two particles come close to each other, the polymer layers overlap and a local increase of the polymer concentration between these particles is present. The resulting increase of the osmotic pressure leads to drift the particles apart. The range of interaction is influenced by the length and the density of the polymer chains. If the layer has a sufficient thickness to overcome the van der Waals forces, the interaction potential of sterically stabilized particles can be described by a simple hard-sphere potential:

$$\Phi_\mathrm{HS}(r) = \begin{cases} \infty, & r < 2a \\ 0, & r > 2a \end{cases} \qquad (2.4)$$

It is zero at inter particles distances larger than the diameter of the particles and infinite for smaller distances. The phase behavior of hard sphere systems is well-investigated and theoretically understood [14–16]. It is purely entropy-driven, which means that the phase behavior of hard spheres is

---

[1]For a dilute solution of N small non-interacting particles or polymers in a Volume V, the osmotic pressure $P_\mathrm{osm} = \frac{N}{V}k_\mathrm{B}T$ can be determined in analogy to an ideal gas.

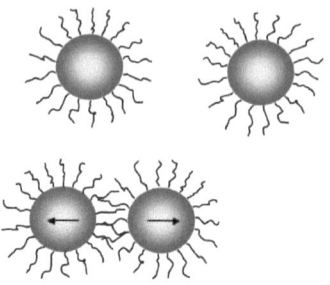

Figure 2.2: Steric stabilization of colloidal particles through a layer of polymer chains on their surface. If two particles come close to each other, the polymer layers overlap and a local increase of the osmotic pressure between these particles leads to a repulsive interaction.

determined by maximizing the entropy. However, increasing polydispersity leads to deviations of the simple phase behavior [17].

In charge stabilized colloidal systems, the particles can be synthesized to carry ionizable surface groups. In a polar solvent with high dielectric constant (e.g. water with $\epsilon_{H_2O} = 80$) the surface groups dissociate due to an energy gain of approximately $100\,k_B T$ per ion as a result of polarization of the molecules. As a consequence, dissociation leads to negatively charged particle surfaces, which are surrounded by their positively charged counter ions. Due to the equal charge of the particles, the entropy tends to distribute them homogeneously over the available space volume. In equilibrium state, the interplay between energy and entropy constitutes an electric double layer as illustrated in Fig. 2.3. The theory of the electric double layer is based on the research of Stern [18], who combined two previous models: a rigid double layer as hypothesized by Helmholtz [19] and a diffusive layer suggested by Gouy and Chapman [20, 21].

Stern suggests that part of the counter ions is adsorbed directly next to the particle's surface, such that the potential across the charged layer decays linearly in analogy to a parallel plate capacitor from Φ_0 at the particle's surface to Φ_S, where Φ_S is known as the Stern potential. Consequently, this layer is referred to as the Stern layer. The remaining ions form a diffusive layer consisting mainly of counter ions and only a small fraction of coions (ions of equal charge as the particle surface). The range of the potential in the diffusive layer decays exponentially and is characterized by the screening length κ^{-1} and the zeta potential ζ. ζ potential is used for estimating the degree of double layer charge and can be measured using electrophoresis methods [22].

2.3.3 The Poisson-Boltzmann equation

To simplify the determination of the electric potential, one initially neglects the adsorption of counter ions at the particle surface and the formation of the Stern layer. The electrostatic forces between the charged particles and the ions counteract thermal motion of the ions. Consequently, the ion concentration in a distance r of the surface can be described by the classical Boltzmann distribution:

$$n(r) = n_0 \exp\left(\frac{-ze\Phi(r)}{k_B T}\right), \tag{2.5}$$

where $n(r)$ is the number density of the counter ions, ze their charge and n_0 their number density at large distances from the particle surface corresponding to a vanishing potential. $\Phi(r)$ is the electrostatic potential as a function of the distance r from the surface. The potential can be determined

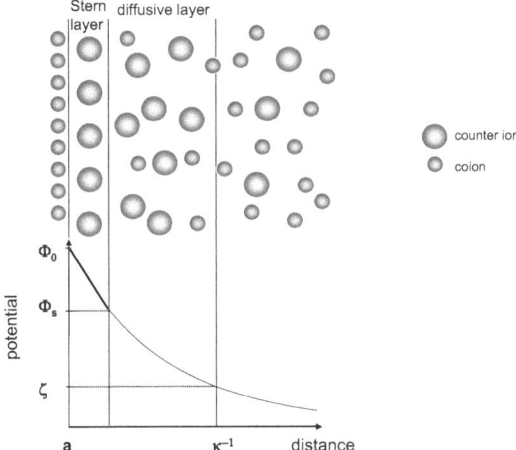

Figure 2.3: Illustration of the double layer model of Stern including the characteristic Stern layer, the diffusive layer and the corresponding potential curve around a charged colloidal particle surrounded by dissociated counter ions and coins.

from the distribution of the entire charge $\rho(r)$ via the Poisson equation:

$$\rho(r) = -\epsilon\epsilon_0 \Delta\Phi(r). \tag{2.6}$$

In the simplest case only counter ions are present as screening ions. In this case both equations 2.5 and 2.6 can be combined leading to the Poisson-Boltzmann equation:

$$\Delta\Phi(r) = -\left(\frac{zen_0}{\epsilon\epsilon_0}\right)\exp\left(\frac{-ze\Phi(r)}{k_BT}\right). \tag{2.7}$$

The interaction can be influenced by adding further ions into a charged system which leads to a screening effect. This more general case has to consider both the negatively and the positively charged ions around the particle. In the simplified case of monovalent ions with the distributions n_+ and n_- of positive and negative ions, respectively, Eq. (2.5) can be modified

$$n_\pm(r) = n_0 \exp\left(\frac{\mp ze\Phi(r)}{k_BT}\right). \tag{2.8}$$

Therefore, the number density of ions as function of the distance is given by:

$$\rho(r) = ze(n_+(r) + n_-(r)) \tag{2.9}$$

The combination of both equations 2.8 and 2.9 leads to the general Poisson-Boltzmann equation considering the presence of additional screening ions:

$$\Delta\Phi(r) = -\left(\frac{2zen_0}{\epsilon\epsilon_0}\right)\sinh\left(\frac{-ze\Phi(r)}{k_BT}\right). \tag{2.10}$$

Eq. (2.10) is an analytically non solvable three dimensional equation. A simplification can be achieved by assuming the electrostatic energy to be much smaller than the thermal energy ($e\Phi(r) \ll k_\mathrm{B}T$). In this case the so called Debye-Hückel approximation [23, 24]

$$n_\pm(r) \approx \left(1 \pm \frac{ze\Phi(r)}{k_\mathrm{B}T}\right) \tag{2.11}$$

can be used. The Debye-Hückel approximation in combination with the Poisson-Boltzmann equation 2.10 leads to the Debye-Hückel equation:

$$\Delta\Phi(r) = \kappa^2 \Phi(r) \quad \text{with} \quad \kappa^2 = \frac{2n_0 z^2 e^2}{\epsilon\epsilon_0 k_\mathrm{B}T}, \tag{2.12}$$

where κ is the Debye screening parameter which is also known as the reciprocal screening length. The screening length is often expressed in terms of the Bjerrum length $\lambda_\mathrm{B} = \frac{e^2}{4\pi\epsilon\epsilon_0 k_\mathrm{B}T}$ as $\kappa = \sqrt{4\pi\lambda_\mathrm{B} n_0}$. The Bjerrum length is the characteristic length, where the Coulomb interaction for monovalent ions becomes comparable to the thermal energy. The solution of the Debye-Hückel equation for monodisperse spherical particles is the Debye-Hückel potential:

$$\Phi(r) = \frac{Ze}{4\pi\epsilon\epsilon_0} \frac{\exp(\kappa a)}{(1+\kappa a)} \frac{\exp(-\kappa r)}{r}. \tag{2.13}$$

Z is the total number of the dissociated surface groups of one particle also denoted as the bare charge. Hence, the surface charge of one colloidal particle is Ze. The electrostatic interaction energy of the two charge distributions yields

$$V(r) = \frac{Z^2 e^2}{4\pi\epsilon\epsilon_0} \left(\frac{\exp(\kappa a)}{(1+\kappa a)}\right)^2 \frac{\exp(-\kappa r)}{r}. \tag{2.14}$$

Here the factor $\frac{\exp(\kappa a)}{(1+\kappa a)}$ is known as the volume exclusion term which considers the fraction of the sample volume which is occupied by colloidal particles and not available to screening ions. Except the volume exclusion term, the Debye-Hückel potential is equivalent to a Yukawa potential.

The Debye-Hückel linearization is valid as long as electrostatic potentials remain low ($e\Phi \ll k_\mathrm{B}T$), independent of the separation between the particles. Its highest value is at the surface (Φ_0), where also the same condition $e\Phi_0 \ll k_\mathrm{B}T$ has to be fulfilled for providing the validity of the repulsive interaction. This condition is in general complied at large distances from the surface. It has been shown, however, that the repulsive potential by means of the Debye-Hückel approximation is also valid at distances in the size range of inter particle separations if the bare charge Z of a colloidal particle is renormalized to the much smaller effective charge Z^* within the Poisson-Boltzmann cell (PBC) model [25] by solving the Poisson-Boltzman equation numerically in a spherical Wigner-Seitz cell. The interaction between charged colloidal particles retain a Debye-Hückel form even when the charge on the particle produces a surface potential which is considerably larger than thermal energies. The exponential screening inherent in the Boltzmann distribution implies that strong potentials are strongly screened by a buildup of counterion charge at small radii. Coupled with the requirement of global charge neutrality, Alexander demonstrated that the Poisson-Boltzmann equation must have a Debye-Hückel like solution in the region near the Wigner-Seitz cell boundary. He also showed that added salt concentration was not strongly renormalized [25]. The use of charge renormalization can be qualitatively explained by the strong adsorption effect of counterions near the surface (Stern layer) which therefore renormalizes downward the particle charge.

In addition, the presence of excess electrolyte concentration ρ_s which leads to further screening of the particles' surface charge, is encountered by an additional term in the screening length in the simplest way by an empirical modification of the Debye-Hückel approximation:

$$\kappa^2 = \frac{e^2}{\epsilon\epsilon_0 k_\mathrm{B}T}\left(nZ^* + \rho_\mathrm{s}\right) \tag{2.15}$$

The latter estimate of the screening parameter (Eq. (2.15)) in combination with the Debye-Hückel potential describes the results of many experiments concerning the properties of charged colloidal suspensions over a wide range of particle number densities and screening electrolyte concentrations [26, 27] and is therefore an excellent approximation for the interaction of charged systems. This kind of interaction is purely repulsive but the presence of attractive contributions in the interaction is also discussed in literature [28].

If particles approach each other to distances of some nm, then the attractive van der Waals interaction becomes relevant. The interplay between attraction and repulsion is described by the classical work of Derjaguin and Landau [29] and Verwey and Overbeek [30] in the 1940s which explains the stability of colloidal suspensions on the basis of an effective potential between pairs of charged particles. This Derjaguin-Landau-Verwey-Overbeek (DLVO) potential consists of a dispersive van der Waals attraction, a steric hard-sphere repulsion and a double layer repulsion of screened Coulomb form. Many attempts to find a theoretical basis for attractions between charged colloids have been reported. However, several experimental investigations with direct measurements of the colloid-colloid pair interaction confirmed that attractive interactions can be neglected and a purely repulsive screened Coulomb form described by the DLVO theory represents the behavior of charged colloids in the best way [28, 31–33].

Having derived an effective interaction potential for charged colloids, the next section presents the connection between interaction and the resulting phase behavior. Hard spheres are one of the most fundamental systems in condensed matter physics. If the hard core potential is taken as infinite when particles inter penetrate and zero when they do not, then the interaction energy is zero and the free energy contains only the entropy term. The hard sphere phase transition is driven purely by configurational entropy, which is constrained by the different possible packings of impenetrable particles at high densities. The phase behavior of hard sphere systems depends on the packing fraction only and is well understood theoretically [14–16]. In other systems like charged stabilized colloids, energy and entropy always compete. Due to the electrostatic repulsion additional control parameters like electrolyte concentration appear to influence the interaction. The resulting phase behavior is more complex and is explained in the next section.

2.4 Phase behavior of charged colloids

The phase behavior of atomic systems depends in general on the temperature T and the pressure p. For charged spheres in aqueous dispersions the effective particle charge Z^*, the particle concentration n and the salt concentration c form the experimental control parameters for the screened Coulomb interaction which also depends on the particle radius a as a constant parameter depending on the analyzed system. Combinations of these take the role of a pressure and an effective temperature in the statistical mechanics calculations [34]. Consequently the phase transition can be induced isothermally by reducing the content of screening electrolyte or increasing the particle concentration. The particle number density can also be specified as the volume fraction Φ in per cent:

$$\Phi = V_\mathrm{p} n = \frac{4}{3}\pi a^3 n, \qquad (2.16)$$

where V_p denotes the particle volume. Depending on these parameters, fluid, crystalline and glass-like[2] phases are formed in colloidal systems. Charged spheres crystallize either bcc or fcc depending on the degree of metastability of the melt, while hard spheres show fcc crystallization only [34, 35].

Fig. 2.4 shows one of the first experimental phase diagrams in dependence of the volume fraction and the electrolyte concentration derived systematically by Small Angle X-ray scattering [36]. Sirota *et al.* observe successively, bcc, fcc and glass-like phases as the volume fraction increased at zero added electrolyte concentration. The bcc phase exists only in a small region of the phase diagram as the electrolyte HCl is added. The excess of screening ions leads to a reduced electrostatic repulsion

---

[2] No gas-liquid phase transition can be observed in charged colloids due to the absence of attractive pair interaction. From the thermodynamical point of view, these systems are the analogon to an overcritical atomic fluid.

and a decrease of the strength of interaction where no crystals of long-range order are observed and the system stays fluid. At large Φ there is a transition from glass to fcc solid to liquid with increasing electrolyte concentration. The solid lines are guide to the eye phase boundaries. Dashed line is the fcc-liquid theoretical phase boundary for a similar point-charge Yukawa system.

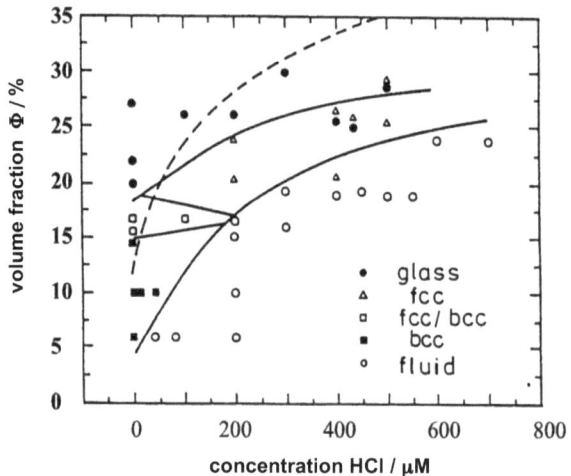

Figure 2.4: Experimental phase diagram in dependence of the volume fraction and the electrolyte concentration derived by Small Angle X-ray scattering [36]. Sirota *et al.* observe successively, bcc, fcc and glass-like phases as the volume fraction increased at zero added electrolyte concentration. The bcc phase exists only in a small region of the phase diagram as the electrolyte HCl is added. The excess of screening ions leads to a reduced electrostatic repulsion and a decrease of the strength of interaction where no crystals of long-range order are observed and the system stays fluid. At large Φ there is a transition from glass to fcc solid to liquid with increasing electrolyte concentration. The solid lines are guide to the eye phase boundaries. Dashed line is the fcc-liquid theoretical phase boundary for a similar point-charge Yukawa system.

Monovoukas *et al.* [37] also observed crystallization at low volume fractions where bcc structure is energetically favored. The bcc crystals melt upon the addition of salt and undergo a structural transformation to an fcc phase upon an increase in Φ. Qualitatively, the experiments show that as the interaction strength is increased, one progresses from the liquid state to a solid which is bcc for soft long-range interactions (low volume fraction and electrolyte concentration) or fcc for hard short-range interactions. The reason therefore lies in the volume fraction Φ dependence of the interaction potential[3]. The particles prefer a densest packing of fcc or hcp structure at short-range interactions and a bcc ordered structure at isotropic long-range interactions. This behavior follows from the fact that the fcc lattice shows higher density of packing than the bcc structure. While the neighbor distance is 3% larger in a fcc lattice in comparison to a bcc lattice, a particle in an fcc lattice has 12 and the one in a bcc lattice only 8 next nearest neighbors, which overcompensates the longer neighbor distance.

Characteristic for the phase behavior of charged colloids is that crystallization occurs at lower volume fractions in contrast to hard sphere systems. The electrostatic repulsion has its maximum value in the deionized state and depends on the distance between the particles only. The isotropic

---
[3]The screening length κ^{-1} is comparable to the average next nearest neighbor distance $\bar{d}_{NN}$ in systems with long-range potentials. While κ^{-1} decreases on a square root dependence with increasing volume fraction, the average next nearest neighbor distance $\bar{d}_{NN}$ decreases on a cube root dependence.

separation leads to a minimization of the electrostatic energy where the most favorable state results in a periodical arrangement of the particles if the electrostatic interaction at their next nearest neighbor distance is larger than the thermal energy $k_B T$. The addition of electrolyte leads to a screening effect and reduction of the interaction. If the electrostatic energy is reduced and of the order of thermal energy, no long-range order can be sustained and the crystal melts.

Phase behavior is an essential part for characterization of a new system. For further analysis detailed investigations of the structure are necessary. To derive informations about the structure of both fluid and solid at equilibrium or non-equilibrium state, a theoretical description of the scattering theory is given in the next section.

2.5 Scattering theory

Since the length scale of colloidal systems is comparable to the wavelength range of visible light, light scattering is a powerful technique for structure analysis in transparent solutions. Application of optical methods are in general limited due to turbidity and multiple scattering at high particle number densities. In addition also the particle distances shrink with increasing volume fraction. In this case more intense radiation sources of lower wavelengths are necessary. Scattering sources of UV light have appropriate wavelengths, but are not usable due to strong absorption effects. For that reason, X-ray scattering was applied at high particle number densities for turbid systems. The resulting difference of the wavelengths between the scattering source (X-ray) and the colloidal inter particle distances (visible light) leads to scattering phenomena at (ultra) small angles which bear experimental challenges.

The theory governing the scattering processes of light or X-rays (LS or XS) is fundamentally the same and is based on the knowledge of classical electrodynamics [38]. Scattering techniques form a valuable set of experimental methods that are very useful for investigating the structure and dynamics of materials. In addition, a combination of different techniques, such as light and X-ray scattering, a large range of length scales from Angstroms to microns is covered. While the scattering contrast in X-ray scattering comes from the electron density difference between the solute and the solvent, in LS it is the refractive index difference. In the present section, light scattering is therefore introduced in detail followed by an outline of differences between both techniques. The remainder of this section is concerned with a brief outline of the description of the melt state in colloids and pure metals.

2.5.1 Light scattering

The theory introduced in this section is usually referred to as the Rayleigh-Gans-Debye scattering theory. An overview is given e.g. by Dhont [39] and for more details the literature of Berne *et al.* [40] is recommended.

Within an ensemble of equal spheres with a fixed radius a, each particle acts as a scattering center for electromagnetic waves. The incident wave is scattered at the particles without changing its wave length or phase. The resulting electromagnetic field corresponds to the sum of all scattered waves in a particular direction. The phase difference of the scattered light depends on the relative position in space of the wave sources as well as on the direction in which the electric field is measured. The scattered intensity contains information concerning the relative position of the scatterers and the internal structure of the particles due to their mesoscopic size. Since the next nearest neighbor distances and the extent of colloidal particles are comparable to the length scale in the order of the wavelength of visible light, light scattering is the dominant measurement technique to study structural properties in mesoscopic systems.

In the special case of an ensemble of colloidal systems, each particle includes various scattering points, such that inter- and intra particle interference of the scattered light is possible. This phenomenon is illustrated in Fig. 2.5. The scattering points within a particle are the electrons which are induced to oscillate by the incident plane wave $\mathbf{E}_i = \mathbf{E}_0 e^{i(\omega t + \mathbf{k}_i \mathbf{r}_j)}$, with the wave vector $\mathbf{k}_i$ of the

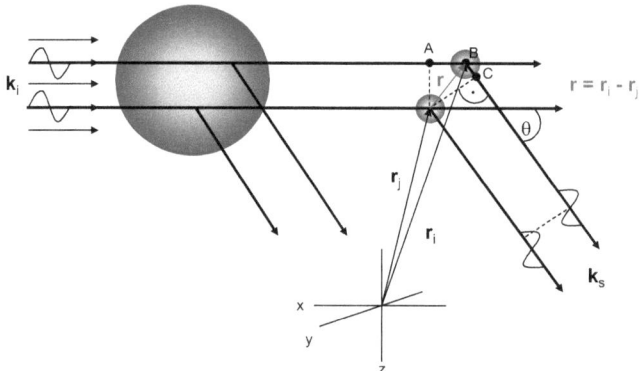

Figure 2.5: Schematic diagram of light scattering in colloidal systems demonstrating two characteristic scattering phenomena: Each colloidal particle of mesoscopic extent (left side) includes various scattering points. In this case, intra particle interference of the scattered light is possible in addition to inter particle interference of light scattered by two different particles (right side).

incident wave, its amplitude $\mathbf{E}_0$ and the coordinate $\mathbf{r}_j$ of the j-th particle, in a way that a dipole moment occurs. The dipole moment of the whole particle is $\boldsymbol{\mu} = \bar{\boldsymbol{\alpha}}\mathbf{E}_i$ with the polarizability tensor $\bar{\boldsymbol{\alpha}}$ of one particle. The phase difference referred to a predefined origin is $\mathbf{k}_i\mathbf{r}_j$. The distance from the origin to the observation point is further assumed to be much larger than any dimension of the scatterer. The so-called far-field approximation [38] purposes to calculate the electric field of the outgoing wave scattered by the j-th particle in the following way:

$$\begin{aligned}
\mathbf{E}_j &= \frac{1}{c^2 R^3}\mathbf{R} \times (\mathbf{R} \times \boldsymbol{\mu})e^{-i\mathbf{k}_s\mathbf{r}_j} \\
&= \underbrace{-\frac{\omega^2}{c^2 R}\alpha}_{\xi_j}\ \underbrace{\mathbf{E}_0 e^{i\omega t}}_{\mathbf{E}_0(t)}\ e^{i(\mathbf{k}_i\mathbf{r}_j - \mathbf{k}_s\mathbf{r}_j)} \\
&= \xi_j \qquad \mathbf{E}_0(t)\ \ e^{i(\mathbf{k}_i\mathbf{r}_j - \mathbf{k}_s\mathbf{r}_j)}
\end{aligned} \qquad (2.17)$$

In this equation $\mathbf{R}$ denotes the distance between the detector and the scattering volume, $\mathbf{k}_s$ the scattered wave vector and c the velocity of light. In the case of spherical particles, the polarizability tensor $\bar{\boldsymbol{\alpha}}$ can be converted into its scalar form a. It is also assumed that the interaction of the electric field with the material occurs without affecting the wavelength. The energy of each photon is the same before and after the scattering process referred to as elastic scattering. Only the direction of the scattered light changes, which is characterized by the scattering angle Θ. That implies the following relation for the scattered wave vector:

$$\mathbf{q} = \mathbf{k}_s - \mathbf{k}_i = \frac{2\pi\nu_s}{\lambda}(\mathbf{e}_s - \mathbf{e}_i) \quad \text{with} \quad |\mathbf{q}| = \frac{4\pi\nu_s}{\lambda}\sin(\Theta/2). \qquad (2.18)$$

Thereby, $\mathbf{e}_s$ and $\mathbf{e}_i$ are the unit vectors in the direction of the scattered and the incident light, and ν_s is the refractive index of the solvent. Eq. (2.18) is explained in figure 2.5 for clarity. The phase difference $\Delta\phi$ of the outgoing waves scattered at two particles which are located at $\mathbf{r}_i$ and $\mathbf{r}_j$ under a scattering angle Θ is equal to $\frac{2\pi\nu_s}{\lambda}\Delta s$ with $\Delta s = \overline{AB} + \overline{BC}$ as it is illustrated in Fig. 2.5 resulting in $\Delta\phi = (\mathbf{r}_i - \mathbf{r}_j)\mathbf{q} = \mathbf{r}_{ij}\mathbf{q}$ for the phase difference. The total electric field strength of scattered light $\mathbf{E}_s$ is the sum over all scattered waves of each scattering point, weighted by the scattering strength $\xi(\mathbf{r})$ of the volume element at $\mathbf{r}$. The scattering strength of a volume element with volume $d\mathbf{r}$ is then

$\xi(\mathbf{r}) d\mathbf{r}$. Replacing the sum over infinitesimal small volume units by an integral yields,

$$\mathbf{E}_s(\mathbf{r},t) = \int_{V_s} e^{i\mathbf{qr}} \mathbf{E}_0 e^{i\omega t} d\mathbf{r} \xi(\mathbf{r}). \tag{2.19}$$

The integration range V_s is the illuminated volume from which scattered light is detected. The derivation of Eq. (2.19) includes several simplified assumptions. Firstly, Eq. (2.19) does not consider the fact that the refractive index of the colloidal particles always differs slightly from that of the surrounding fluid. A phase difference occurs between the incident field which traverses only the fluid part of the sample and the one which traverses a colloidal particle.

Multiple scattering is also neglected in the simple approximation of Eq. (2.19) because higher order scattering events are negligible, if a small fraction of incident light is scattered. This is the case in diluted colloidal systems. Typical volume fractions for charged colloidal suspensions are in the range of only a few percent.

Eq. (2.19) makes no distinction between the interference of light scattered from volume elements within a single colloidal particle and from distinct particles. Since the scattering strength is only non-zero within the colloidal particle, Eq. (2.19) can be written as a sum of electric field strengths scattered by single particles:

$$\mathbf{E}_s(\mathbf{r},t) = \sum_{j=1}^{N} \int_{V_j} e^{i\mathbf{qr}} \mathbf{E}_0 e^{i\omega t} d\mathbf{r} \xi(\mathbf{r}). \tag{2.20}$$

Here, the integration range is given by V_j which is the volume of j-th single particle. Its position coordinates of the center of mass are denoted by $\mathbf{r}_j$ and $\mathbf{r}'$ is a vector from the center of mass to an other point within the colloidal particle. With $\mathbf{r} = \mathbf{r}_j + \mathbf{r}'$, the scattered field strength is given by

$$\mathbf{E}_s(\mathbf{r},t) = \sum_{j=1}^{N} e^{i\mathbf{qr}_j} \int_{V_j} e^{i\mathbf{qr}'} \mathbf{E}_0 e^{i\omega t} d\mathbf{r}' \xi(\mathbf{r}'). \tag{2.21}$$

The first term of Eq. (2.21) containing the position coordinates of the particles' mass centers $\mathbf{r}_j$ describes the interference of light from different colloidal particles. The integral on the other hand, describes interference of light scattered within single particles. It is also defined as the scattering amplitude $b(\mathbf{q})$ of the j-th particle. $\xi(\mathbf{r}) = \frac{\epsilon(\mathbf{r})-\epsilon_f}{\epsilon_f}$ denotes the scattering strength [39]. $\xi(\mathbf{r})$ depends entirely on the optical properties of the colloidal system, which are described by the dielectric matrix. For isotropic particles in the fluid, the dielectric matrix is a constant as a function of the position $\mathbf{r}$. For positions outside the particle, the dielectric constant of the system is equal to that of the fluid ϵ_f which is independent of the position $\mathbf{r}$ for a homogeneous fluid. Scattering from the solvent is neglected here, so that only scattered intensity due to inhomogeneities in the dielectric constant caused by the present particles are considered.

The electric field strength is not accessible in scattering experiments. What is measured instead, is the instantaneous intensity $i(\mathbf{q},t)$. In static light scattering experiments, the ensemble averaged properties of density fluctuations are measured. The ensemble average can be approximated by its time average in practice by detecting the intensity over a time scale that is much larger than the time needed for Brownian motion [39]:

$$I(\mathbf{q}) = \langle i(\mathbf{q},t) \rangle = \frac{1}{2}\sqrt{\frac{\epsilon_f}{\mu_0}} \langle \mathbf{E}_s(\mathbf{r},t) \mathbf{E}_s^*(\mathbf{r},t) \rangle \tag{2.22}$$

$$= \underbrace{\frac{V_s}{R^2}\sqrt{\frac{\epsilon_f}{\mu_0}} E_0^2 \,\overline{\rho}\, (\hat{\mathbf{n}}_0 \hat{\mathbf{n}}_s)^2}_{I^*} \frac{1}{N} \left\langle \sum_{i=1}^{N} \sum_{j=1}^{N} b_i(\mathbf{q}) b_j(\mathbf{q}) e^{i\mathbf{q}(\mathbf{r}_i - \mathbf{r}_j)} \right\rangle, \tag{2.23}$$

where $\langle\,\rangle$ denotes the ensemble average, $\hat{\mathbf{n}}_0 = \frac{\mathbf{E}_i}{|\mathbf{E}_i|}$ is the polarization direction of the incident and $\hat{\mathbf{n}}_s = \frac{\mathbf{E}_s}{|\mathbf{E}_s|}$ the one of the scattered light. $\bar{\rho}$ expresses the density of the colloidal particles. For monodisperse spherical particles which show no orientational correlation, the scattering amplitudes can be taken outside the ensemble average, which delivers the following expression for the intensity:

$$I(\mathbf{q}) = \underbrace{I^*b(q=0)^2}_{=\ I_0} \underbrace{\left|\frac{b(q)}{b(q=0)}\right|^2}_{P(q)} \underbrace{\frac{1}{N}\left\langle\sum_{i=1}^{N}\sum_{j=1}^{N} e^{i\mathbf{q}(\mathbf{r}_i-\mathbf{r}_j)}\right\rangle}_{S(\mathbf{q})} \qquad (2.24)$$

Here $P(q)$ denotes the particle form factor and $S(\mathbf{q})$ the structure factor.

The particle form factor $P(q)$ depends on the particle size and shape. It describes the interference of the waves scattered at different volume elements within single particles. For optically homogeneous and monodisperse particles, $P(q)$ is given by

$$P(q) = 9\left[\frac{qa\cos(qa) - \sin(qa)}{(qa)^3}\right]^2, \qquad (2.25)$$

with a denoting the radius of the particles. Eq. (2.25) includes the simplified assumption that the difference of the refractive index between the particles and the fluid is small and further that the particles are small in comparison to the wave length of the radiation. This is also known as the so called Rayleigh-Gans approximation. Polydispersity or deviations from spherical shape and the corresponding influence of particle orientation can also be treated within this approximation [41]. For larger particles and those with radial variations in the refractive index, the Mie scattering theory yields the exact calculations [42].

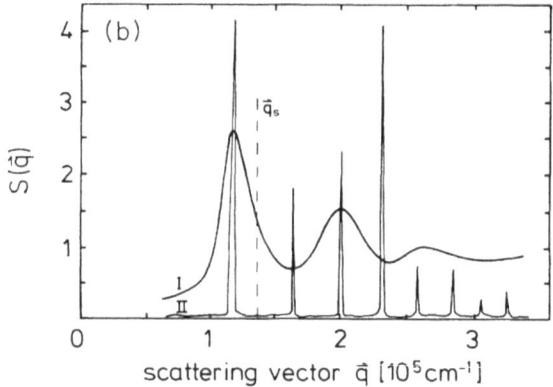

Figure 2.6: Structure factor of the fluid (I) and the crystalline state (II) of a charged polystyrene suspension at a volume fraction of $\Phi = 0.0032$. The difference in the phase behavior results in different salt concentrations. High counter ion concentration $c_s = 1.3\mu mol/l$ (I) induces a fluid state due to the screening effect and the lower ones $c_s < 0.2\mu mol/l$ a crystalline phase. (figure taken from [43])

The inter particle structure factor $S(q)$ describes the interference of the electric fields of waves scattered at different particles. $S(q)$ reflects the structure of the colloidal system and is defined as the Fourier transformation of the pair-correlation function $g(r)$:

$$S(q) = 1 + (4\pi/\rho)\cdot\int r\left[g(r) - 1\right]\cdot\sin(qr)\,dr, \qquad (2.26)$$

where ρ is the particle number density.

For an ideal crystal $S(q)$ takes the form of a delta function. In case of finite crystal size, broadening of Bragg peaks is observed. If a crystal is very small or if stacking faults are present, a considerable finite broadening of the Bragg peaks is possible so that intensity is also present at scattering vectors which do not fulfill the Bragg condition [44].

Fig. 2.6 shows the structure factor of the fluid (I) and the crystalline state (II) of a charged polystyrene suspension at a volume fraction of $\Phi = 0.0032$. The difference in the phase behavior results in different salt concentrations. High counter ion concentration $c_s = 1.3 \mu mol/l$ (I) induces a fluid state due to the screening effect and the lower ones $c_s < 0.2 \mu mol/l$ a crystalline phase [43].

The fluid state is characterized by the existence of short-range order instead of long-range correlations present in the crystalline state. The short-range order can be described by the probability of finding the center of a particle at a given distance from the center of another particle. For short distances, this is related to how the particles are packed together. Further away, the surrounding particles are diffusely arranged, and so for large distances, the probability of finding two spheres with a given separation is essentially constant. In that case, it is related to the density of the system. The mathematical expression for this probability is the pair correlation function $g(\mathbf{r}_1, \mathbf{r}_2)$. The pair correlation function of isotropic systems depends on the absolute value of the distance only: $g(\mathbf{r}_1, \mathbf{r}_2) \to g(r)$ and is normalized in a way that in the case of completely uncorrelated particles it converges against 1 at large values of r.

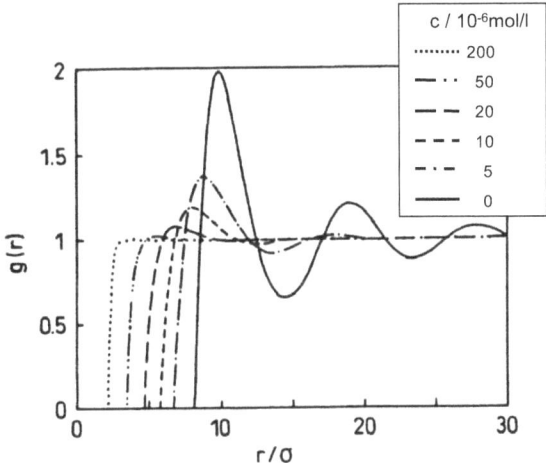

Figure 2.7: Pair correlation functions of charged polysterene particles with a radius of $a = 84nm$, an effective charge of $Z^* = 438$ and a volume fraction of $\Phi = 6.2 \cdot 10^{-4}$ in aqueous solution at different electrolyte concentrations $c = 0, 5, 10, 20, 50, 200 \cdot 10^{-6} mol/l$. The dotted line corresponds to the highest electrolyte concentration and therefore lowest interaction energy where the pair correlation function describes a gas-like order without any intercorrelations between the particles. The solid line corresponds to an electrolyte-free suspension with the highest interaction energy, which shows a highly ordered fluid state [41].

In Fig. 2.7 pair correlation functions are shown for a charged polystyrene particles with a radius of $a = 84nm$, an effective charge of $Z^* = 438$ and a volume fraction of $\Phi = 6.2 \cdot 10^{-4}$ in aqueous solution at different electrolyte concentrations $c = 0, 5, 10, 20, 50, 200 \cdot 10^{-6} mol/l$. The dotted line corresponds to the highest electrolyte concentration and therefore lowest interaction energy where the pair correlation function indicates a gas-like order without any intercorrelations between the particles. The solid line corresponds to electrolyte-free suspension with the highest interaction energy, which shows a highly ordered fluid state [41].

2.5.2 X-ray diffraction and light scattering

X-ray and light scattering are interference phenomena described by the general equation

$$E_s = E_0 \sum_i \xi_i \exp\left[i\left(\mathbf{q}\mathbf{r}_i\right)\right], \tag{2.27}$$

where E_s and E_0 are the electric field strengths of the scattered and incident radiation, ξ_i is the fraction of the radiation scattered by the i-th scatterer, $\mathbf{r}_i$ is a vector from an arbitrary origin to the element and $\mathbf{q} = \mathbf{k}_s - \mathbf{k}_i$ is the difference between unit vectors of the incident and scattered rays. The main difference between X-ray and light scattering is related to the property of ξ_i. It is associated with electron densities for X-rays and with the polarizabilities or refractive indices for light scattering. Due to the higher energies of X-rays in comparison to visible light photons, chemical bondings are hardly influenced by X-ray scattering and polarization independent.

The theoretical background of LS and XS is quite similar, mainly dealing with the angular dependence of time-averaged scattered intensity $I(q)$ versus the scattering angle, in terms of the scattering vector q. For a dilute isotropic suspension of N particles per Volume V, the total scattered intensity is given by the product of the intensity $I_0(q)$ of scattering of a single particle and the structure factor $S(q)$ [45, 46]:

$$I(q) = \frac{N}{V} \cdot I_0(q) \cdot S(q), \tag{2.28}$$

where q is the length of the scattering vector, given by

$$q = \frac{4\pi}{\lambda} \sin\left(\Theta/2\right). \tag{2.29}$$

The particle form factor $P(q)$ that is normalized to unity at $q = 0$ is linked with $I_0(q)$ by:

$$I_0(q) = V_p^2 \left(\overline{\rho} - \rho_m\right)^2 P(q), \tag{2.30}$$

where V_p denotes the volume of a single particle, $\overline{\rho}$ is the scattering length density of the particles and ρ_m is the respective quantity of the solvent. With the volume fraction given by $\Phi = (N/V) \cdot V_p$, it follows that $I(q=0)/\Phi = V_p \left(\overline{\rho} - \rho_m\right)^2$ where $\overline{\rho} - \rho_m$ is the excess scattering length density or the contrast of the particles in the respective solvent. With the weight concentration C (weight per unit volume) and the molecular weight M, the scattered intensity denotes:

$$I(q) = KCMP(q) \cdot S(q). \tag{2.31}$$

Structural information is revealed in $P(q)$ and $S(q)$ independently of the scattering technique. The difference of the scattered intensity $I(q)$ for LS and XS is hidden in the optical parameter K. For light scattering, K is given by [47, 48]:

$$K \equiv \frac{4\pi^2 n_0 \left(\frac{dn}{dC}\right)^2}{\lambda^4 N_A}, \tag{2.32}$$

where n_0 is the refractive index of the solvent, (dn/dC) is the refractive index increment, λ the wavelength of the light and N_A denotes the Avogadro's number. The refractive index increment reflects how much the refractive index of a solution varies for a given increment in concentration of solute C.

In case of XS the constant K follows from the number of scattering units per unit mass of the solute [48, 49]:

$$K \equiv \frac{e^4 \left[N(Z-Z_0)/VC\right]^2}{(mc_0^2)^2 N_A}. \tag{2.33}$$

This equation is valid for an ensemble of N particles in a volume V at dilute solute concentration C with Z being the number of electrons per particle, Z_0 the number of electrons per particle volume of the suspending medium, m the electron mass, c_0 the velocity of light and the electron charge e.

Scattering experiments using electromagnetic radiation differs in the wavelength, with visible light in the 350-700nm range and X-rays varying from ca. 0.01-0.2nm. There is no intrinsic limitation over the wavelength in the electromagnetic spectrum from very short wavelength X-rays (in the γ-ray region) to the infrared region, except for the range between long-wavelength X-rays of about 3nm and the vacuum ultraviolet region where the radiation is intensely absorbed by all materials and therefore rendering the scattering by such probing radiation difficult to deal with. Both equations, (2.33) and (2.32), point out the difference that the scattering contrast in XS comes from the electron density difference between the solute and the solvent, while in LS it is the refractive index difference [4].

2.5.3 Model for short-range order

The simplest models for the short-range order analyze the stability of small structural units consisting of a few atoms. It is assumed that such types of structural units are predominantly formed in the melt that are energetically favorable. The development of short-range order in metallic melts is schematically outlined in Fig. 2.8. At its beginning dimers (b) of atoms with a distance $r_{\min}$ are formed. For Lennard-Jones systems, the interaction potential exhibits a potential minimum at $r_{\min}$ that corresponds to the favorable distance between the atoms under equilibrium state. If one atom is attached to the dimers such that triangular clusters (c) are build up, the number of bondings per atom is increased, which reduces the energy. A further reduction of the energy can be obtained by attachment of one more atom on top of the triangle under formation of a regular tetrahedron (d). A tetrahedron is the simplest structural unit with a comparatively high packing density and low energy. The process of aggregation continues until larger polytetrahedral units consisting of several tetrahedra are formed (e-f). 20 tetrahedra form an icosahedron (g) under a slight deformation of the originally regular tetrahedra [50]. By further attachment of atoms, even larger polytetrahedral aggregates such as the dodecahedron (h) are formed [51]. Both, the icosahedron and the dodecahedron are platonic bodies [52] of five-fold symmetry which is incompatible with the translational invariance of crystals. It is impossible to fill space by packing of exclusively clusters of such a symmetry. Therefore, no long-range order of five-fold symmetry exists in nature [53].

Till the early 1950s, it was assumed that the short-range order in undercooled metallic melts resembles that of the corresponding crystalline phases, which nucleate from the melt. Such similarity of the short-range order in the undercooled melt and solid state implies a low energy of the interface between the nucleus of the solid phase and the liquid and therefore a low activation threshold for nucleation. In contrast, Turnbull observed large undercoolings in metallic melts [1] and raised the similarity of the short-range order in the solid and the undercooled melt to question.

2.5.4 Local order in undercooled liquids: metals *vs.* colloids

The undercooling of liquids below their melting temperatures is a well known subject which was extensively studied by Turnbull using emulsion techniques [1] and drop-tube processing [54], by Bardenheuer and Bleckmann using melt-fluxing methods [55] and by Herlach applying containerless processing methods through levitation [56]. Due to small length scales in atomic systems, the microscopic process that suppress the transition to the more favorable stable crystalline phase cannot be observed directly and a full understanding of local ordering in undercooled melts is still missing.

---

[4] In neutron scattering experiments, the scattering contrast comes from deuterized parts between the solute and the solvent.

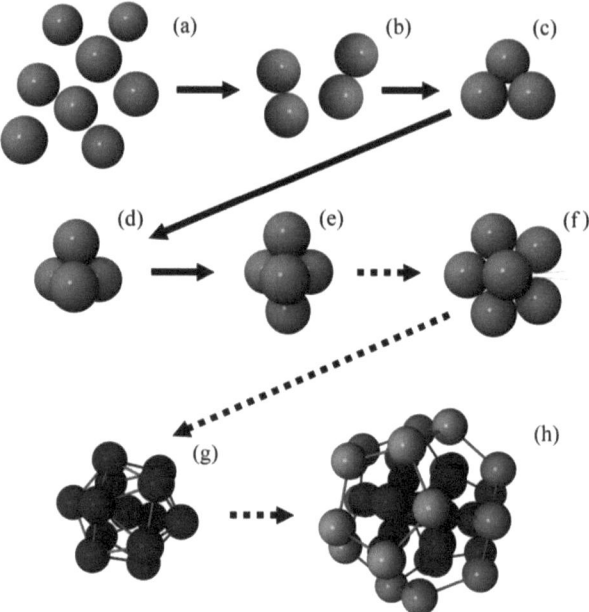

Figure 2.8: Development of short-range order in metallic melts. At the beginning dimers (b) of atoms with a distance r_{min} are formed. Increasing the number of bondings per atom reduce the energy, if one atom is attached to the dimers such that triangular clusters (c) are build up. A significant reduction of the energy can be obtained by attachment of one more atom on top of the triangle under formation of a regular tetrahedron (d). A tetrahedron is the simplest structural unit with a comparatively high packing density and low energy. The process of aggregation continues until larger polytetrahedral units consisting of several tetrahedra are formed (e-f). 20 tetrahedra form an icosahedron (g) under a slight deformation of the originally regular tetrahedra [50]. By further attachments of atoms, even larger polytetrahedral aggregates such as the dodecahedron (h) are formed [51].

In 1952, Frank [2] postulated that the short-range order in undercooled metallic melts is based upon icosahedral aggregates independent of the structure of the corresponding solid phases. Frank showed that the energy of an icosahedral aggregate of 13 Lennard-Jones atoms is 8% lower than that of crystallographic clusters with the same number of atoms (e.g. aggregates of fcc or hcp structure). Frank gave the explanation for Turnbull's large undercoolings in the melts investigated. The energy calculations for an icosahedral cluster include no explicit assumptions on the degree of undercooling or metastability. Frank's hypothesis does not exclude the preference of icosahedral type of short-range order in stable liquids above the melting point.

Experimental evidence of icosahedral short-range order in liquid metals came from X-ray diffraction, neutron scattering and X-ray absorption spectroscopy investigations [57–60]. In the case of semi-metals like Sb, investigations of the short-range order showed a preference for clusters with a cubic structure [61]. Here a pronounced degree of covalent bonds has to be taken into account. Therefore, the assumption of non-directional bonds, which is the basis for the prediction of an icosahedral short-range order, appears unjustified for Sb.

In atomic systems there are no experiments concerning microscopic structure providing information that go beyond averaged properties as the pair distribution function $g(r)$ or the structure factor $S(q)$.

One possibility to clarify the question of the predominant short-range order in undercooled liquids is to use computer simulations. Steinhardt, Nelson and Ronchetti [62,63] studied bond-orientational order in molecular-dynamics simulations of undercooled liquids and concluded that the local order is predominantly icosahedral in the undercooled state. Sachdev and Nelson [64] developed a statistical mechanics theory for icosahedral short-range order in dense fluids and conclude that icosahedral order should dominate in undercooled liquids. Also in computer simulations of metastable Lennard-Jones (LJ) liquids an icosahedral short-range order was found that increases with rising undercooling [65]. A fraction of 61% of atoms are organized within an icosahedral environment that is expected to be even larger in the bulk of real LJ liquids like metals [66]. In simulations using hard sphere potentials, on the other hand, groups of 13 atoms do not form icosahedral units [67]. Moreover, for layers of fluid hard sphere systems close to a crystalline substrate no icosahedral short-range order was detected [68].

An alternative way to analyze the local order is the use of model systems. Charged colloids are ideal model systems due to their large length and relaxation time scale. They provide the possibility to tune the Debye-Hückel potential from a long-range coulombic interaction (with the screening parameter $\kappa = 0$) to a hard-sphere like one ($\kappa \to \infty$). Colloidal suspensions show a phase behavior which is analogous to that of atomic systems with a main advantage that also real-space investigations of the formation of local order are possible. Gasser investigated the local order of weakly charged colloids by laser scanning confocal microscopy. The local order of the particles was analyzed with a bond order parameter method. The investigations revealed that even the smallest nuclei had a fcc and hcp like character with a high disorder especially close to the solid-liquid interface [69]. Investigations on the local order in the fluid state, however, revealed a bond order parameter that indicate a surrounding that is close to a fragment or a whole icosahedron for more than 50% of all particles [70]. In addition, the icosahedral character of the local order increased with volume fraction due to the higher strength of interaction.

2.6 Kinetics of crystallization

The basic processes in phase transition of first order are nucleation, growth and eventually ripening or coarsening. These processes are not sharply distinguished, but in fact overlap during the evolution from a metastable (undercooled or supersaturated) to a condensed phase. During the first part of the last century, a number of semi-empirical descriptions for the kinetics of nucleation and growth have been developed, mainly by adapting models of the condensation of liquid drops from the vapor phase. Nucleation phenomena are characterized within the framework of classical nucleation theory (CNT) based on the work of Volmer and Weber [71]. This model has been extended by Becker and Döring [72], Zeldovich [73] as well as Turnbull and Fischer [74]. Nucleation without preferential nucleation sites is called homogeneous nucleation that occurs spontaneously and randomly, but it requires undercooling of the medium. CNT predicts an exponential dependence of homogeneous nucleation rate densities J on the free energy barrier for nucleus formation. Spherical growing nuclei must overcome an energy barrier in order to keep on growing under energy profit. Solidification is also influenced by growth after heterogeneous nucleation induced either at the container walls or impurities in the bulk [75, 76]. Reaction controlled growth velocities follow a Wilson-Frenkel type behavior [77, 78].

2.6.1 Classical nucleation theory

Metastable liquids tend to undergo a phase transformation due to its high driving force for nucleation. In a liquid the atoms follow random motions driven by temperature. These density fluctuations lead to agglomeration of particles in a statistical way, forming solid like aggregates (clusters). When assuming that such a cluster is spherical with radius r, the energy balance for its formation is given by the sum of two contributions, a surface term $\Delta G_{\text{surface}}$ and a volume term ΔG_{volume}:

$$\Delta G(r) = \underbrace{4\pi r^2 \gamma}_{\Delta G_{\text{surface}}} + \underbrace{\frac{4}{3}\pi r^3 n \cdot \Delta \mu}_{\Delta G_{\text{volume}}}, \tag{2.34}$$

where γ is the solid-liquid interfacial energy, n the particle number density and $\Delta \mu$ the difference of chemical potential between the metastable melt and stable solid state. As work has to be done to create the interface ($\gamma > 0$), the surface term is always positive. The volume term depends on the chemical potential difference $\Delta \mu$ between the fluid and solid phase. It serves as a measure for undercooling and is always negative for a system below the melting point. Eq. (2.34) is often

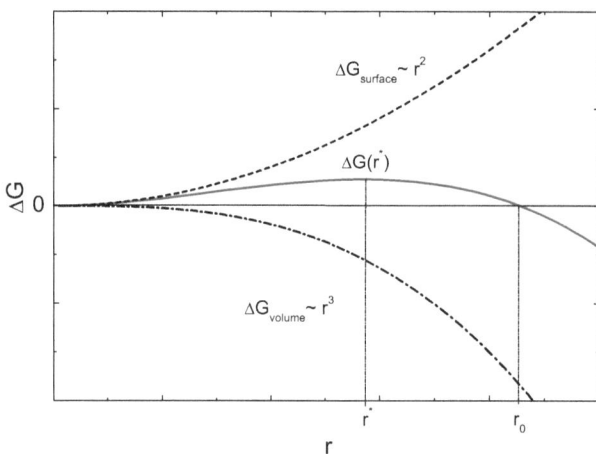

Figure 2.9: Energy balance of a solid-like cluster in a melt as a function of the cluster radius r. Nucleation is influenced of two competing terms of the Gibb's free energy: the surface term $\Delta G_{\text{surface}}$ and the volume term ΔG_{volume}. ΔG^* is the activation energy for the formation of a nucleus of critical size r^*. The nucleus is able to grow under energy gain after overcoming the activation energy ΔG^* and eventually becomes stable at $r0 = 1.5r^*$.

referred to as the capillarity approximation [79] which is plotted in Fig. 2.9. A characteristic feature of this function is a maximum with the height ΔG^* at a critical radius r^*. This value corresponds to an activation barrier for crystallization. For clusters smaller than r^* in size, the interface term dominates so that the resulting unstable cluster has a high likelihood to decay spontaneously. For a large enough radius the volume term begins to dominate the (unfavorable) surface term, the nucleus gain energy by further growing and eventually becomes stable at $r_0 = 1.5r^*$.

Differentiation and solving of the extreme value problem of Eq. (2.34) leads to the critical nucleus size r^*:

$$r^* = -\frac{2\gamma}{n\Delta\mu} \tag{2.35}$$

with the activation barrier

$$\Delta G^* = \Delta G(r^*) = \frac{16\pi}{3} \frac{\gamma^3}{(n\Delta\mu)^2}. \tag{2.36}$$

The particle number in the critical nucleus can be determined as

$$i^* = \frac{4}{3}\pi r^{*3} n = \frac{32\pi\gamma^3}{3n^2 |\Delta\mu|^3}. \tag{2.37}$$

The number of nuclei arising per time and volume unit, the nucleation rate density, has the form [72]:

$$J = J_0 \exp\left(\frac{-\Delta G^*}{k_\mathrm{B} T}\right), \qquad (2.38)$$

with J_0 as the kinetic prefactor. Eq. (2.38) shows that the nucleation rate density is proportional to the thermodynamic probability of fluctuations leading to the formation of a critical cluster and a dynamical factor which is also called attachment or condensation factor describing the attachment of single particles to the nucleus.

The presented treatment of homogeneous nucleation is based on phenomenological thermodynamics that is a continuum theory for macroscopic systems. Therefore it is questionable if its description remains valid for small nuclei consisting of a small particle number. The macroscopic interface energy γ is also defined only for systems of planar interface in equilibrium state. However, nucleation in a metastable system is a non equilibrium process. Additionally it should be mentioned that this theory describes the nucleation at constant pressure and temperature. This assumption is obviously not fulfilled in metallic melts but is always considered for simplification. In colloidal systems, in contrast, the crystallization can be considered as an isothermal process[5].

2.6.2 Kinetics of homogeneous nucleation

The kinetics of homogeneous nucleation process is based on the process of growth and shrinkage of clusters which is described by an attachment or separation of single particles using rate equations [74, 79, 81]:

$$J = K^+ \cdot Z \cdot n_\mathrm{nucleus}. \qquad (2.39)$$

The steady state nucleation rate density is determined by the condensation rate K^+, by Z the Zeldovich factor which takes the width of nucleation rate barrier into account and by the concentration of critical clusters in the melt n_nucleus which is described using a Boltzmann distribution:

$$n_\mathrm{nucleus} = n_\mathrm{liq} \exp\left(-\frac{\Delta G^*}{k_\mathrm{B} T}\right), \qquad (2.40)$$

where n_liq is the particle number density in the colloidal fluid.

The Zeldovich factor is defined as the second derivation of the nucleation barrier and reads as follows:

$$Z = \sqrt{\frac{(\partial^2 \Delta G/\partial i^2)_{i=i^*}}{2\pi k_\mathrm{B} T}} = \sqrt{\frac{\Delta G^*}{3\pi k_\mathrm{B} T i^{*2}}}. \qquad (2.41)$$

In adaption to colloidal systems Eq. (2.41) reads as follows [5]:

$$Z = \frac{1}{8\pi} \frac{n_\mathrm{xtal} \Delta \mu^2}{\gamma^{3/2}} \frac{1}{\sqrt{k_\mathrm{B} T}} \qquad (2.42)$$

where n_xtal is the particle number density in the colloidal nucleus. At the critical size of a nucleus the condensation (K^+) and the evaporation rate (K^-) coincide with the maximum of the nucleation barrier height. But a particle condensed at the nucleus has to pass the fluid crystal interface and has to cross an activation energy ΔG_A which is given by the difference in the free enthalpy between the activated cluster $\Delta G(i+1)$ and the inactivated $\Delta G(i)$. The condensation rate for a critical nucleus is therefore described by a Boltzmann distribution and follows a form similar to the theory of absolute reaction rates:

$$K^+ = 4i^{*2/3} f \exp\left(-\frac{\Delta G_A}{k_\mathrm{B} T}\right) = 4 n_\mathrm{xtal}^{2/3} \left(\frac{4\pi}{3}\right)^{2/3} \left(\frac{2\gamma}{n_\mathrm{xtal} \Delta \mu}\right)^2 f \exp\left(-\frac{\Delta G_A}{k_\mathrm{B} T}\right), \qquad (2.43)$$

---

[5]The latent heat ΔH releasing during the crystallization of a colloidal suspension is in the order of $k_\mathrm{B} T$ per particle. For a typical particle number density of $n = 10^{19} m^{-3}$ in a colloidal system, Williams [80] et al. estimated the latent heat to $\Delta H = 4.8 kcal/mol$. It leads to an increase in temperature of $\Delta T = n N_A c_w \Delta H = 8 \cdot 10^{-8} K$, with the Avogadro's number N_A and c_w the specific heat of water. Therefore, crystallization in colloidal systems can be considered as an isothermal process.

where f is the vibration frequency. Assuming that the activation energy can be expressed by the diffusion coefficient D and the next nearest particle distance d_{NN} in the fluid:

$$f \exp\left(-\frac{\Delta G_A}{k_B T}\right) = \frac{6D}{d_{NN}^2}, \tag{2.44}$$

the attachment rate K^+ reads as follows:

$$K^+ = 24 n_{xtal}^{2/3} \left(\frac{4\pi}{3}\right)^{2/3} \left(\frac{2\gamma}{n_{xtal}\Delta\mu}\right)^2 \frac{D}{d_{NN}^2}. \tag{2.45}$$

By substituting Eqs. (2.40), (2.42) and (2.45) in Eq. (2.39) one arrives at the final result for the nucleation rate density:

$$J = 12 \left(\frac{4}{3}\right)^{2/3} \pi^{-1/3} \sqrt{\frac{\gamma}{k_B T}} \frac{D}{d_{NN}^2} n^{2/3} \exp\left(-\Delta G^*/k_B T\right). \tag{2.46}$$

Eq. (2.46) takes into account that in monodisperse charged colloidal systems under fully deionized conditions, the particle number density of the crystal and the fluid are equal [5].

2.6.3 Transient nucleation

The derivation of the nucleation rate described in the previous section is based on the assumption of a stationary nucleation process with a stationary cluster distribution in the melt. It is well known that the nucleation process can be a non steady state in case of rapid changes of states compared with internal relaxation times. In this case, elapsed time is observed before the first critical nucleus is formed. While in a first state of nucleation, the nucleation rate is fairly small, it increases with time and finally approaches a constant value. A steady-state of the nucleation rate is achieved after a transient time, also known as an induction time or time lag, where the cluster distribution achieves its instantaneous equilibrium. During rapid cooling of metals and alloys, for example, the experimental time as given by the high cooling rate can be short compared with the characteristic time to establish the steady state condition so that sufficient time does not remain to establish the dynamic equilibrium in the disturbed cluster system. This leads to deviations from the stationary cluster distribution and to transient effects.

The first theoretical description for the time dependent nucleation rate was given by Zeldovich [73]. Today, the most commonly used equation was suggested by Collins [84] and Kashchiev [85], who assumed a distribution of relaxation times:

$$J(t) = J_{ss} \left[1 + 2 \sum_{m=1}^{\infty} (-1)^m \exp\left(-\frac{m^2 t}{t_i}\right)\right]. \tag{2.47}$$

Here, J_{ss} denotes the steady-state nucleation rate corresponding to the time independent J in Eq. (2.38) and the induction time, t_i, is given by [85]:

$$t_i = \frac{4}{\pi^3 K^+ Z^2}, \tag{2.48}$$

where K^+ is the condensation rate and Z the Zeldovich factor. Eq. (2.48) represents the result of the transient time following the theory of Kashchiev. Whereas Kashchiev and Collins evaluate the same expression in Eq. (2.47) for the time dependent nucleation rate, both give a slightly different expression for the transient time[6]. By substituting Eqs. (2.42) and (2.43) in Eq. (2.48), the following expression for the transient time is given for charged colloidal systems:

$$t_i = \frac{\gamma k_B T d_{NN}^2}{\frac{3}{8} \left(\frac{4}{3}\right)^{\frac{2}{3}} \pi^{\frac{5}{3}} D n^{\frac{2}{3}} \Delta\mu^2}. \tag{2.49}$$

---
[6]Collins evaluate the following expression for the transient time: $t_i = \frac{1}{\pi^2 K^+ Z^2}$ [84].

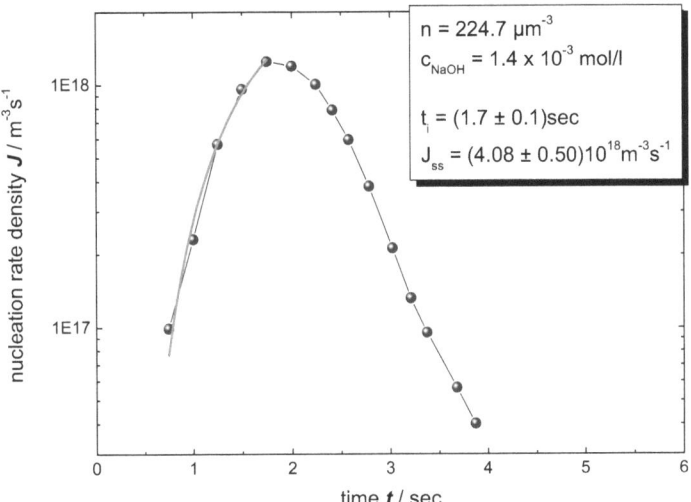

Figure 2.10: Evaluation of the steady-state nucleation rate I_{ss} using Eq. (2.47) as a two-parameter fit for experimentally determined time dependent nucleation rate density for a silica suspension of spheres with a 77nm diameter ($n = 224.7 \mu m^{-3}$ and $c_{NaOH}= 1.4 mmol/l$). If no free volume remains available for crystal growth, the nucleation rate collapses necessarily [82, 83] without achieving the steady state nucleation rate which is necessary for the evaluation of nucleation properties within the framework of the classical nucleation theory.

As the nucleation proceeds and the nuclei start growing beyond the critical size, the melt volume shrinks during further crystal growth. If no free volume remains available for crystal growth, the nucleation rate collapses necessarily. This phenomenon is demonstrated in Fig. 2.10. While the nucleation rate achieves the maximum value of $J_{max} = 1.23 \cdot 10^{18} m^{-3} s^{-1}$ at about 2 sec and collapses due the unavailable free volume, an obviously higher value for the steady state nucleation rate of $J_{ss} = 4.08 \cdot 10^{18} m^{-3} s^{-1}$ is observable. The calculations of the transient nucleation rate as presented in Eq. (2.47) are based on the assumption of homogeneous nucleation, but can readily be transferred to the case of heterogeneous nucleation [86].

2.6.4 Diffusion

The behavior of colloidal particles is influenced by irregular motion caused by thermal excitations due to collisions with the molecules of the solvent. This motion is also called Brownian motion which is characterized by the typical time τ_B. The Brownian time τ_B is a measure for the average time in which the velocity of a Brownian particle relaxes to thermal equilibrium with the surrounding fluid. The Brownian relaxation time is in the order of 10^{-8}s. For large times $t \gg \tau_B$, an isolated Brownian particle moves diffusively through the solvent with a well defined mean-square displacement which varies linearly with time:

$$\langle \Delta r(t)^2 \rangle = 6Dt. \quad (2.50)$$

The proportional constant D in Eq. (2.50) denotes the diffusion coefficient. In a system of colloidal particles one can distinguish between self-diffusion (D_S according to the motion of a single particle) and collective diffusion (D_C according to the motion of density fluctuations caused by collective motion of the particles). The self-diffusion coefficient of an isolated sphere with radius a, suspended in a solvent with viscosity η is determined by the Stokes-Einstein self-diffusion coefficient:

$$D_0 = \frac{k_B T}{6\pi\eta a}. \tag{2.51}$$

The Stokes-Einstein diffusion coefficient D_0 depends on macroscopic properties only. However, for high particle concentrations or in the case of electrostatic repulsion, interactions between the particles become significant and lead to complex behavior of the diffusion coefficient. Self-diffusion phenomena are distinguished at different time scales between short-time D_S^S and long-time D_S^L behavior. The transition between both regimes is characterized by the time τ_i, the time one colloidal particle needs to diffuse a mean separation distance between suspended particles. Short-time diffusion is observed for $t \ll \tau_i$ and long time time diffusion for $t \gg \tau_i$.[7] In a concentrated suspension of colloidal particles embedded in a solvent, the long-time self-diffusion coefficient of the colloidal particles D_S^L, is significantly smaller than the short-time diffusion constant D_S^S. Whereas the latter results from random kicks of the solvent and is determined in terms of the solvent friction and the temperature, the long-time self-diffusion is strongly affected by the repulsive inter particle interactions. Furthermore, both quantities depend on hydrodynamic interactions mediated by the solvent. These complicated interactions can be safely ignored if their range, characterized by a hydrodynamic radius, is much smaller than the range of the inter particle interaction. This is the case, for example, in highly charged colloidal suspensions which already show a well-pronounced structure even for very small packing fractions [87].

2.6.5 Crystal growth

After introducing the formation of nuclei considering transient effects and the diffusive behavior of charged colloids, the next step of crystallization has to be described: the growth mechanism of a nuclei. Independently of each other, Wilson [77] and Frenkel [78] developed a simple theoretical model to explain crystal growth.

Wilson and Frenkel assumed that the attachment of single particles from the metastable melt to a crystalline phase takes place via diffusion processes over an activation threshold, which is built up by the energy potential of the surrounding particles. The rate with which the particles undergo a transition from the liquid into the crystalline lattice is given by the following expression:

$$\alpha = l \cdot f \cdot \exp\left(-\frac{E_D}{k_B T}\right). \tag{2.52}$$

This equation includes a characteristic length for particle diffusion l, a vibration frequency f per unit of area and the activation barrier E_D for diffusion.

At the same time, particles, which dissociate in the fluid phase, counteract the crystallization process. As the free enthalpy per particle, the chemical potential μ_f in the fluid phase is larger than in the crystalline phase, the dissociation rate for the backward reaction is reduced by the factor $\exp\left(-\frac{\Delta\mu}{k_B T}\right)$ in comparison to the crystallization rate:

$$\beta = l \cdot f \cdot \exp\left(-\frac{E_D}{k_B T}\right) \cdot \exp\left(-\frac{\Delta\mu}{k_B T}\right). \tag{2.53}$$

with $\Delta\mu = \mu_f - \mu_s$ the chemical potential difference between the fluid and solid phase.

The idea of Wilson and Frenkel for growth of crystals is shown in Fig. 2.11. Here the attachment of one particle of fluid-like ordered phase (grey spheres) into a vacant place in the crystalline phase (blue spheres) is demonstrated assuming a monoatomic interface between the fluid and the crystal. The behavior of the free energy together with diffusion barrier E_D and the chemical potential difference $\Delta\mu$ is displayed above the scheme. The arrow illustrates the idea of the Wilson-Frenkel theory: The

---

[7]In colloidal systems this time is in the order of $\tau_i \approx 10^{-4}s - 10^{-3}s$.

barrier for attachment of a particle to the surface of the crystalline framework depends on the self diffusion coefficient D_S of a particle in the fluid phase.

The difference between Eq. (2.52) and (2.53) gives the resulting growth velocity in terms of the Wilson-Frenkel equation:

$$v = \alpha - \beta = l \cdot f \cdot \exp\left(-\frac{E_D}{k_B T}\right)\left[1 - \exp\left(-\frac{\Delta \mu}{k_B T}\right)\right] \qquad (2.54)$$

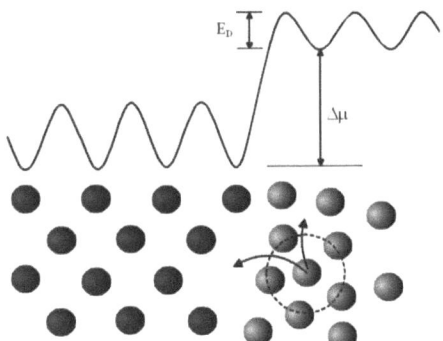

Figure 2.11: Attachment of one particle of fluid-like ordered phase (grey spheres) into a vacant place in the crystalline phase (blue spheres) assuming a monoatomic interface between the fluid and the crystal. The behavior of the free energy together with diffusion barrier E_D and the chemical potential difference $\Delta \mu$ is displayed above the scheme. The arrow illustrates the idea of the Wilson-Frenkel theory: The barrier for attachment of a particle to the surface of the crystalline framework depends on the self diffusion coefficient D_S of a particle in the fluid phase.

Assuming that the vibration frequency can be expressed using the self diffusion coefficient $D_S = l^2 f \exp\left(-\frac{E_D}{k_B T}\right)$, the Wilson-Frenkel equation 2.54 reads:

$$v = \frac{D_S}{l}\left[1 - \exp\left(-\frac{\Delta \mu}{k_B T}\right)\right]. \qquad (2.55)$$

At times $t \gg \tau_B$ the dominant physical process for a Brownian particle is the increasing difficulty to diffuse out of the cage formed by its neighbors. Therefore, the long time self-diffusion coefficient D_S^L is the most convenient choice for calculating the growth velocity in Eq. (2.55). The characteristic length scale for particle diffusion is equated with the next nearest neighbor distance: $l = d_{NN} \approx n^{-1/3}$. The derivation of Eq. (2.55) includes the simple assumption of a monoatomic interface between the fluid and the crystal.

Based on investigations of Palberg [88], Stipp [89] and Broughton [90] indications of an extended interface layer were observed. It is assumed that particles in the adjacent liquid layer of thickness d_I attach the crystal surface at a rate proportional to D_S^L/d_{NN}^2. A factor $f_0 < 1$ is included to account for the fact that some of these collisions are ineffective for crystallization. The resulting growth velocity reads as follows:

$$v = \underbrace{f_0 d_I \cdot \frac{D_S^L}{d_{NN}^2}}_{v_\infty}\left[1 - \exp\left(-\frac{\Delta \mu}{k_B T}\right)\right], \qquad (2.56)$$

where the prefactor v_∞ denotes the limiting velocity at large undercoolings. Growth is driven by the difference in the chemical potential difference. At large $\Delta \mu$, however, it is limited by a kinetic process leading to a finite and constant maximum growth velocity. To adapt this formalism to colloidal systems, an assumption for the chemical potential difference is necessary. A simple approximation was first suggested by Aastuen et al. [91] that $\Delta\mu$ increases linearly with $n - n_\text{F}$, once the freezing point n_F is exceeded:

$$\Delta\mu = B\, k_\text{B} T\, \frac{n - n_\text{F}}{n_\text{F}}, \qquad (2.57)$$

where B is a fitting parameter that needs to be determined from experiment.

2.7 Theory of elasticity

Under appropriate conditions of particle charge, particle number density and electrolyte concentration structural ordering is observed in colloidal systems (sec. 2.4). Since the crystals are stabilized by purely repulsive Coulomb interactions (sec. 2.3.3), they form a classical analog to Wigner crystallization predicted for electrons [92]. One of the earliest realizations of such a colloidal crystal was in Tipula iridescent virus [93]. Due to the fact that lattice distances lie in the range of about 0.5 μm iridescence (opalescence) originating from Bragg scattering of visible light was observable.

The two basic properties which characterize a crystalline solid as opposed to a liquid are the existence of structural order and rigidity. Bragg scattering is useful to establish the existence of the solid state in crystals (sec. 2.5.1), but does not reveal information about the interaction which is responsible for the crystallization. Rigidity on the other side is evidenced by a finite shear modulus. Elastic properties are measures for the internal forces and an important instrument to study the interaction. This section is concerned with the classical theory of elasticity to establish a basis for important elastic quantities with special reference to the theory of Landau and Lifschitz [94].

An important quantity in theory of elasticity is the stress tensor as a response of strain in a deformed solid. The assumptions for describing elasticity in atomic systems can be adapted to systems of larger length scale like colloids. Joanny [95] derived the equations of motion for a coupled water-interacting polystyrene system and found there were low-frequency propagating shear waves. The first to observe this phenomenon was Pieranski $et\ al.$ [96], who induced standing shear waves in a driven cylinder of the colloidal crystal and detected their presence by the amplitude of the motion of Kossel lines.

2.7.1 Elasticity of crystalline systems

External forces lead to stress fields in a solid which in turn result in strain (deformation). For a quantitative description of this effect a displacement vector, a strain tensor and a stress tensor are introduced in the following.

The displacement field $\mathbf{u}$ defines the displacement of any particle in a strained body from its original (unstrained) position $\mathbf{x}$ to the position $\mathbf{x}'$ in the strained state:

$$\mathbf{u} = \mathbf{x}' - \mathbf{x}, \qquad (2.58)$$

which is a position dependent vector with its components u_j. If the displacement is small, the strain can be treated by conventional linear elasticity theory. In that case, the linearized strain tensor is defined as

$$e_{ij} = \frac{1}{2}\left(\frac{\partial u_j}{\partial x_i} + \frac{\partial u_i}{\partial x_j}\right) \qquad (2.59)$$

The strain components depend on the derivatives of the n-dimensional system x_i with $i = 1, \ldots n$. This means that the strain tensor is an $n \times n$-matrix, which is symmetric ($e_{ij} = e_{ji}$). The symmetry of the stress tensor follows from the law of conservation of angular momenta in the classical case, i.e., in the absence of internal moments, surface moments and mass moments. Otherwise, the external force would lead to rotations and not to deformation. We obtain the normal strain as the diagonal elements of the strain tensor and the shear strains are contained in the rest of the tensor.

Deformations caused by external forces, induce an internal stress in a solid which is described by the stress tensor σ_{ij}. Elasticity theory links the strain in a volume element to the forces acting on this element:

$$\sigma_{ij} = \sum_{k,l} c_{ijkl} \cdot e_{kl}. \tag{2.60}$$

This relation includes the fourth-rank tensor c_{ijkl}, which is also known as the elasticity tensor. Eq. (2.60) is also denoted as the generalized Hooke's law which extends the one dimensional concept of a spring constant to three-dimensional linear anisotropic solids.

The work of straining of an elastic solid is stored as energy in it. The energy stored per unit volume is called strain energy density. For elastic strain with small deformation, the strain energy density is equivalent to the free energy F:

$$F = F_0 + \sum_{i,j} \frac{1}{2} \sigma_{ij} e_{ij}, \tag{2.61}$$

with the constant free energy F_0 of the unstrained material in equilibrium state. Referring to Hooke's law, Eq. (2.60), the energy density can also be written as

$$F = F_0 + \sum_{i,j,k,l} \frac{1}{2} c_{ijkl} e_{ij} e_{kl}. \tag{2.62}$$

Due to the symmetry of the strain tensor, any permutation of the indices leaves the product $e_{ij} \cdot e_{kl}$ invariant leading to the following index symmetry conditions $c_{ijkl} = c_{jikl} = c_{ijlk} = c_{klij}$, which reduces the 81 elements of the fourth-rank tensor to 21 independent components. Depending on the crystal system, the number of independent elastic components can further reduced. For cubic crystals the only three independent components in three dimensions are

$$C_{11} = c_{xxxx} = c_{yyyy} = c_{zzzz},$$
$$C_{12} = c_{xxyy} = c_{yyzz} = c_{zzxx},$$
$$C_{44} = c_{xyxy} = c_{yzyz} = c_{zxzx}. \tag{2.63}$$

In this case the so called Voigt's notation [97] of the elasticity tensor is introduced with the reduction of the four indices to twofold indexed elements C_{ij}. The elasticity tensor for cubic systems can be converted to a simplified form following Voigt's notation [97]:

$$\begin{pmatrix} C_{11} & C_{12} & C_{12} & & & \\ C_{12} & C_{11} & C_{12} & & 0 & \\ C_{12} & C_{12} & C_{11} & & & \\ & & & C_{44} & & \\ & 0 & & & C_{44} & \\ & & & & & C_{44} \end{pmatrix}. \tag{2.64}$$

The quantities C_{ij} are called the elastic stiffness constants (or the elastic moduli). The inverse elements of C_{ij} are called the elastic compliance (or the elastic constants). In the case of isotropic systems, the elasticity tensor is further simplified by the relation $C_{44} = \frac{C_{11} - C_{12}}{2}$.

The stiffness constants are related to the shear modulus in dependence of the direction in which shearing is applied. One simple relation between the shear modulus and the stiffness constants arises for shearing of crystal planes perpendicular to the main axes. This case is illustrated in Fig. 2.12, where a body centered cubic unit cell with a lattice constant g is shown in the unsheared state on the left side and in the sheared state on the right side. In the latter case a tangential force is exerted in the [100] direction. These conditions lead to $G = C_{44}$.

Elastic behavior of solids is first of all influenced by the interaction between the particles. Considering the binding energy as a sum over all two-body interactions, the relation between G and the

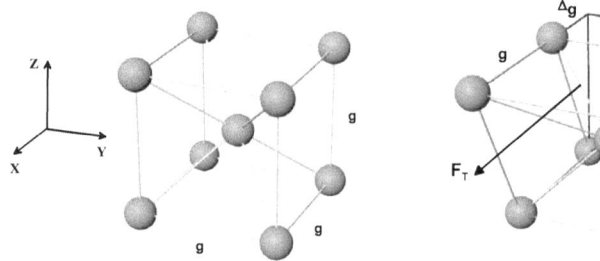

Figure 2.12: Body centered cubic unit cell with a lattice constant g is shown in the unsheared (left side) and in the sheared state (right side). In the latter case a tangential force $\mathbf{F}_T$ is exerted in the [100] direction (figure taken from [6]).

interaction energy V can be determined. The relation between G and V for the body centered cubic (bcc) lattice at shearing in [100] direction, can be derived as [98, 99]:

$$\nu_{bcc}C_{44} = \nu_{bcc}G_{bcc} = \frac{4}{9}\left(\frac{\partial^2 V}{\partial r^2}\right)_{r=d} d^2 + \frac{8}{9}\left(\frac{\partial V}{\partial r}\right)_{r=d} d, \qquad (2.65)$$

where d is the next-nearest neighbor distance and $\nu_{bcc} = \frac{4}{3\sqrt{3}}d^3$ the volume of the unit cell.

The elastic behavior of a system depends on the applied shear direction and the condition of anisotropy must be considered. The anisotropy of the elastic properties in cubic systems is defined by the anisotropy factor $A = \frac{2C_{44}}{C_{11}-C_{12}}$. $A = 1$ represents isotropic behavior and a deviation from 1 determines the strength of the elastic anisotropy.

Elastic anisotropy for a bcc and a fcc single crystal is demonstrated in Fig. 2.13 and is well known in literature [97, 100–102]. In the case of polycrystalline systems, an empirically determined relation can be used to describe the macroscopic shear modulus in averaging over randomly oriented crystallites or local environs

$$G = f \cdot C_{44}, \qquad (2.66)$$

with the so-called morphology factor f. For polycrystalline systems a value of $f = 0.5$ is encountered in most cases and will also be used in this thesis. In general for a bcc crystal G is largest in [111] direction and minimum for shear in [100] direction. For charged stabilized colloidal solids these values have been computed for single crystals together with averaged values for randomly oriented polycrystalline systems [26].

2.7.2 Elasticity of colloidal systems

In the previous section 2.7.1 it was shown that the shear modulus is proportional to the second derivative of the potential with respect to the particle displacement (Eq. (2.65)). Using a screened Coulomb potential for the interaction in a charged colloidal system following relation to the shear modulus of a polycrystalline system with bcc structure is derived [103]:

$$G_{bcc} = f\frac{4}{9}n\left(V(r)\right)_{r=d}\kappa^2 d^2. \qquad (2.67)$$

The parameters influencing G_{bcc} are the particle number density n, the excess counter ion concentration c and and the effective charge Z^*. The Coulomb interaction between two particles in colloidal

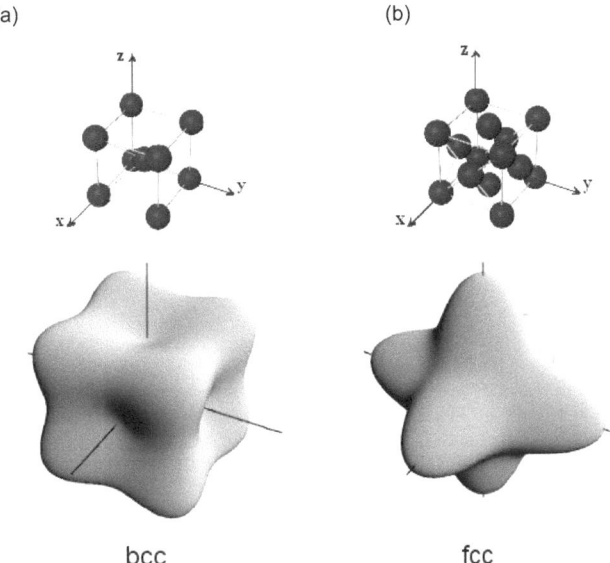

Figure 2.13: Anisotropy of the shear modulus for a bcc a) and a fcc b) single crystal: the distance between the origin of the co-ordinate system and the surface yields the value of the shear modulus in the given direction (figure taken from [6]).

systems can be estimated to be in the order of about 10eV.[8] This is comparable to the interaction in an atomic solid. On the other hand, there is a large difference in the values of the shear modulus between atomic and colloidal systems, which comes from the difference in particle number density[9]. The shear modulus G of colloids is about 12 orders of magnitudes lower in comparison to atomic systems.

In addition to scattering techniques, also elastic measurements are a complementary technique for analyzing the phase behavior. The presence of standing shear waves indicates the crystalline phase. They are absent in the fluid or even in the coexistence phase. The shear modulus is also sensitive to bcc and fcc phase transitions [104].

Characteristic for colloidal systems is the presence of a viscous solvent which causes high friction for the moving particles. This effect strongly dampens the lattice vibrations. Since the solvent is incompressible, its volume conservation must be taken into account. The crystal lattice in contrast, can change its lattice constant and hence its volume periodically as it is shown in Fig. 2.14 for longitudinal waves. Pieranski demonstrated in a simple estimate that the friction due to Stoke's law leads to an overdamping of longitudinal waves, so that only transversal waves are possible in a colloidal system [105, 106]. Longitudinal modes are strongly overdamped since the incompressibility of the fluid necessitates a counterflow in the case of density variation.

---

[8] Coulomb interaction $Z^2 e^2/\epsilon r$ between two spheres with a charge $\sim 1000e$ and a separation of $\sim 1000 nm$ in a medium with a dielectric constant of ~ 100.

[9] Particle number density in atomic solids is $\sim 1 \mathring{A}^{-3}$ in contrast to colloidal crystals $\sim 1 \mu m^{-3}$.

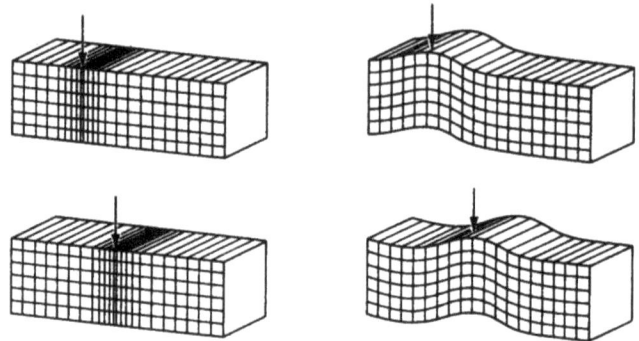

Figure 2.14: Propagation of longitudinal (left) and transversal (right) waves in a homogeneous medium of elastic behavior. In the case of longitudinal waves, local compression of volume elements with changing lattice constants can be observed. In contrast, propagating transversal waves conserve the crystal volume and the lattice constant in the medium. The propagation direction is in both cases from left to right (figure taken from [107]).

2.7.3 Shear waves in colloidal solids

Hydrodynamics describes the motion of colloidal solids assuming a two component system which consists of particles and the solvent. For an isotropic, homogeneous and incompressible system, the equation of motion for the particles can be derived as [95, 96, 108, 109]

$$n \cdot \xi \left(\mathbf{v} - \frac{\partial \mathbf{s}}{\partial t} \right) + G \cdot \Delta \mathbf{s} = \mathbf{0}, \tag{2.68}$$

and the one for the solute medium as (Navier-Stokes equation):

$$\rho \frac{\partial \mathbf{v}}{\partial t} = n \cdot \xi \left(\frac{\partial \mathbf{s}}{\partial t} - \mathbf{v} \right) + \eta \cdot \Delta \mathbf{v}. \tag{2.69}$$

The equations depend on the particle number density n, frictional coefficient $\xi = 6\pi\eta a$ of one particle with radius a (Stoke's friction), viscosity of the solvent η, the velocity of the solvent $\mathbf{v}$, displacement of particles $\mathbf{s}$, shear modulus G and the density of the solvent ρ.

The first term of Eq. (2.68), $n \cdot \xi \left(\mathbf{v} - \frac{\partial \mathbf{s}}{\partial t} \right)$, describes the friction between particles and the fluid (Stokes friction). The second term, $G \cdot \Delta \mathbf{s}$, describes the elastic force of embedded particles in a crystalline framework. For an isolated particle this term can be neglected. The term, $\rho \frac{\partial \mathbf{v}}{\partial t}$, of Eq. (2.69) represents the Archimedes force, which is exerted by the accelerated fluid on a particle in absence of friction. Pressure is not considered in this derivation since only transverse modes are taken into account. Both equations ((2.68) and Eq. (2.69)) in combination result to:

$$\rho \cdot \frac{\partial^2 \mathbf{s}}{\partial t^2} - \left(\eta + \frac{\rho G}{n\xi} \right) \Delta \frac{\partial \mathbf{s}}{\partial t} - G \cdot \Delta \mathbf{s} + \frac{\eta G}{n\xi} \cdot \Delta^2 \mathbf{s} = 0 \tag{2.70}$$

The second term in Eq. (2.70) can be reduced to $\eta \Delta \frac{\partial \mathbf{s}}{\partial t}$ due to the friction between the colloidal particles and the solvent compared to the shear modulus $\frac{\rho G}{n\xi\eta} \ll 1$ [95]. The last term, $\frac{\eta G}{n\xi} \cdot \Delta^2 \mathbf{s}$, in Eq. (2.70) can also be omitted, because it corresponds to a mode with a penetration depth which is in the order of the particle distance and lower. In this case no oscillation can be transferred from one particle to another and no oscillation can be contributed to the response of the whole sample. The equation of motion is then given by:

$$\rho \cdot \frac{\partial^2 \mathbf{s}}{\partial t^2} - \eta \Delta \frac{\partial \mathbf{s}}{\partial t} - G \cdot \Delta \mathbf{s} = 0 \tag{2.71}$$

Using cylindrical coordinates solutions of the second order differential equation 2.71 have the form of harmonic oscillations:
$$\mathbf{s}(r,z,t) = \mathbf{s}(r,z)e^{i\omega t}. \quad (2.72)$$
Solutions of equation 2.71 for standing shear waves with the corresponding boundary conditions are Bessel functions of order one, J_1. The boundary conditions are determined by the sample cell geometry (radius R) and the filling level (height H) in the sample cell. Initially, a partially filled sample cell is used. The resulting conditions for the deflection of the shear wave at the bottom and on the lateral walls of the cylindrical cell is $s(r,0) = s(R,z) = 0$. Additionally, the surface of the suspension has to be stress-free due to the incompressibility and volume conservation must be taken into account: $\left(\frac{\partial s(r,z)}{\partial z}\right)_{z=H} = 0$.

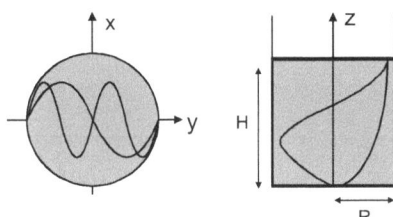

Figure 2.15: Resonant modes of a partially filled cylindrical cell of radius R. The filling level is denoted by H. In each case, the first two resonant modes ($jm = 10; 20$) are illustrated in dependence on the distance of the torsional center r (left side) and in dependence of the height z (right side).

Normal modes of vibration are simple harmonic oscillations of a system where any motion can be exactly expressed as a superposition of normal modes, which are independent, i.e., one mode cannot be expressed as a linear combination of the others. The frequencies of the normal modes[10] of a system are known as resonant frequencies. In mode j, all N particles of the system oscillate at the same frequency, ω_j.

The normal modes which satisfy Eq. (2.71) under the mentioned boundary conditions are:
$$s_{j,m}(r,z) = J_1\left(\mu_j \frac{r}{R}\right) \sin\left\{\left(m + \frac{1}{2}\right)\pi \frac{z}{H}\right\} e^{i\omega t}, \quad (2.73)$$
where μ_j[11] are zero's of the first order Bessel function J_1. The indices j, m denote the resonance modes ($j = 0, 1, 2 \ldots$ are the main modes and $m = 0, 1, 2 \ldots$ are the submodes). The resonance modes s_{10} and s_{20} in dependence of r and z are shown in Fig. 2.15.

The frequencies of the resonance modes are:
$$\omega_{j,m} = \sqrt{\frac{G}{\rho}\frac{1}{R}}\sqrt{\mu_j^2 + \left(m + \frac{1}{2}\right)^2 \pi^2 \alpha^2}, \quad (2.74)$$
where $\alpha = \frac{R}{H}$ is the aspect ratio or the geometrical factor of the cylindrical cell.

In the case of a fully filled cell, an additional boundary condition has to be considered: $s(R,H) = 0$. The solution of the equation of motion and the corresponding resonance frequency are:
$$s_{j,m}(r,z) = J_1\left(\mu_j \frac{r}{R}\right) \sin\left\{(m+1)\pi \frac{z}{H}\right\} e^{i\omega t}, \quad (2.75)$$

---

[10] The normal mode spectrum of a 3-dimensional system of N atoms contains $3N - 6$ normal modes ($3N - 5$ for linear molecules in 3D). In general, the number of modes is the system's total number of degrees of freedom minus the number of degrees of freedom that correspond to pure rigid body motion (rotation or translation)

[11] $\mu_1 = 3.83$, $\mu_2 = 7.01$, $\mu_3 = 10.17$, $\mu_4 = 13.22$

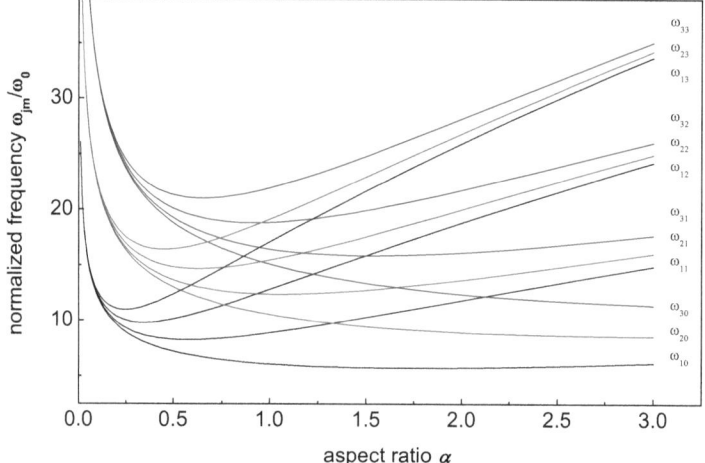

Figure 2.16: Normalized frequencies of normal modes ω_{jm}/ω_0 with $\omega_0 = \sqrt{G/\rho}V^{-1/3}$ at constant sample volume as a function of the aspect ratio $\alpha = \frac{R}{H}$.

$$\omega_{j,m} = \sqrt{\frac{G}{\rho}}\frac{1}{R}\sqrt{\mu_j^2 + (m+1)^2 \pi^2 \alpha^2}. \qquad (2.76)$$

The importance of the aspect ratio α is demonstrated in Fig. 2.16, where frequencies of normal modes are plotted as a function of the aspect ratio α. In addition the frequencies are normalized with $\omega_0 = \sqrt{G/\rho}V^{-1/3}$ and $V = \pi R^2 H$ as the volume of the cell.

For a given value of j, the main resonance splits in subresonances. At low values of α (≈ 0) the subresonances overlap and can hardly be distinguished. At higher values for α (≈ 3) the sequence of the submodes is difficult to analyze. The initial order depending of the values of j at low α, is progressively lost under subsequent overlapping of subresonances. To exclude these disadvantages, the optimum value for the lower frequencies of normal modes occur for $\alpha = 0.5$ at a given volume of cylindrical samples.

In this section, the resonant modes and the resonant frequency for a colloidal system in a defined cylindrical geometry of the experimental sample cell was derived. At a given height H and radius R of the cell and the solvent's density ρ, the shear modulus is directly related to the resonant frequency through 2.74 and 2.76 for different boundary conditions. The resonance frequency ω_{jm} can be experimentally determined using Torsional Resonance Spectroscopy (TRS). This measurement technique is described in sec. 3.2.3.

Chapter 3
Experimental techniques

Pioneering work to develop a multifunctional light scattering experiment combining different techniques was performed by Phan *et al.* [110]. They investigated systematically viscoelastic properties of hard sphere colloidal crystals from resonance detected with dynamic light scattering. In the case of charged colloidal systems, a multipurpose device in combination with an advanced preparational technique was developed by H. J. Schöpe [111, 112]. This technique was adapted and build up at DLR Cologne by P. Wette including further improvements with respect to static and dynamic light scattering experiments.

Due to the fragility of colloidal crystals, it is necessary to have an experimental setup which enables simultaneous investigations of static, dynamic and elastic properties without disturbing the structure. This setup combines static light scattering (SLS), dynamic light scattering (DLS) and torsional resonance spectroscopy (TRS). While SLS and DLS include measurements of angular distributions of scattered intensity, TRS is an angular independent opto-mechanical method for the determination of elastic properties. All three experiments can be performed quasi-simultaneously on the same mechanically undisturbed sample. In addition, also a microscopy experiment is implemented in the multipurpose setup and time-resolved measurements of crystal growth are possible.

Due to the fact that for colloids, particle sizes and distances are comparable with wavelengths in the range of visible light, light scattering is a useful technique for structure analysis in such systems. However, optical methods are in general limited due to turbidity and multiple scattering effects at high particle number densities. In addition, with increasing particle concentrations also the inter particle distances become smaller. To avoid multiple scattering and to guarantee an appropriate wavelength, radiation sources of smaller wavelengths have to be used. Sources of wavelengths in the ultraviolet regime, however, do not come into question due to their strong absorption in colloidal systems. Therefore, X-ray scattering adapted to colloidal systems is an appropriate scattering technique for turbid systems and small lattice spacings. The enlargement of the accessible range of momentum transfer q by using X-rays implies scattering at small angles. The corresponding Ultra Small Angle X-ray Scattering (USAXS) experiments were performed at the soft matter beamline BW4 at HASYLAB (DESY) in Hamburg.

3.1 Interaction control in charge stabilized colloids

The central unit of the experimental setup employed for the systematic investigations on structure, dynamics and elastic properties is the preparation circuit which consists of a closed condition tubing system. This setup was first proposed by Wittig and Palberg [113] and further developed by Wette concerning improvements to avoid flow effects in the sample [114].

The improved preparation circuit enables the control of important parameters like the particle number density or the sodium hydroxide concentration in a precise way. Simultaneously, it guarantees a fast and reproducible method to prepare colloidal suspensions. The basis for the realization of the named conditions is a computer controlled closed tubing system as shown in Figure 3.1. The tubing consists of Teflon® tubes except of the peristaltic pump. Due to squeezing, the pump needs a more

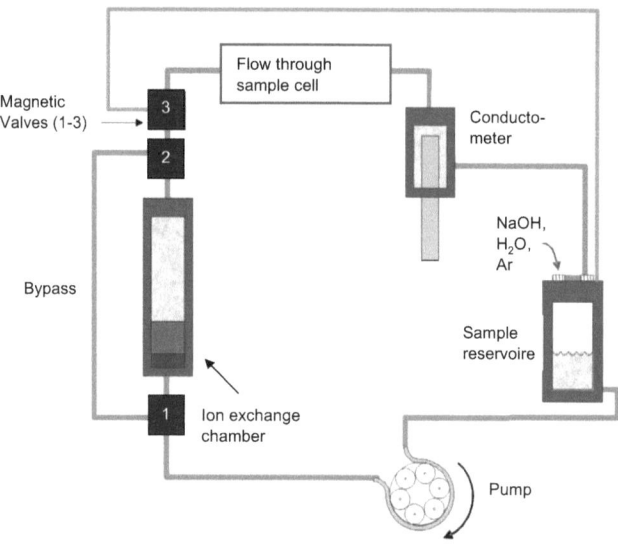

Figure 3.1: Scheme of the computer controlled preparation circuit: The suspension is cycled peristaltically from the sample reservoir to the sample cell and the conductivity measurement device. For deionization it passes additionally through the ion exchange chamber.

resistant and elastic tube of Tygon®. The tubing connects different components. These comprise a mixed bed ion exchange column, a reservoir to add solvent or NaOH solution, the sample cell (for microscopy, light scattering or USAXS measurements) and a conductivity sensor to control the particle number density n for deionized suspensions and the sodium hydroxide concentration c_{NaOH}. The suspension is driven through the preparation circuit by a peristaltic pump under an inert argon atmosphere to avoid contamination with air-borne CO_2. Both solvent and NaOH are added using computer controlled dosimeters (Titronic Universal, Schott AG, Germany). For deionization, the suspension is pumped through an ion exchange chamber instead of the bypass.

Cycling the suspension peristaltically through the closed tubing system has proven suitable to guarantee gradient free, homogeneous shear melts and reproducible adjustment of control parameters within short times. Furthermore the solids may easily be shear molten by pumping the suspension through the preparation circuit. This is possible due to their extremely low shear modulus which is on the order of a few Pa only [115]. Whereas the energies of interaction are of the same order as in atomic systems the low shear rigidity is caused by the low particle densities. As compared with metals, e.g. iron has shear modulus of 81.6 GPa [116]. Consequently metastable colloidal melts can be prepared from well characterized solids without gradients in the control parameters. In addition, computer controlled electromagnetic valves are used to stop the flow instantaneously, which marks the start of a crystallization experiment.

In the following, four measurement techniques of colloidal physics and the involved data handling is introduced. In sec. 3.2.1 and 3.2.2 the static and dynamic light scattering is described for characterization structural and dynamical properties of colloidal systems. An important role for characterization play measurements of elastic properties adapted to colloids using Torsional Resonance Spectroscopy. The determination of the shear modulus reveal a property which is directly related to the interaction potential as a measure of its strength. Sec. 3.2.3 describes this measurement technique followed by the description of microscopy experiments in sec. 3.3 and the USAXS technique in sec. 3.4.

3.2 Multi purpose light scattering instrument

The basic construction of the multipurpose device, in particular the double-arm goniometer for the simultaneous measurements of static, dynamic and elastic properties, was developed by H. J. Schöpe [111, 112]. This technique was adapted and build up at DLR Cologne by P. Wette including further developments to improve the setup particularly with regard to the automatization of the measurements. A sketch of the setup is shown in Fig. 3.2 and a photograph of the central goniometer unit in Fig. 3.3.

With this instrument static light scattering, dynamic light scattering and elasticity measurements can be performed simultaneously on the same sample of well controlled interaction. Simultaneity is crucial, as it circumvents transfer of the sample which conveys the danger of disturbing its sensitive morphology or even shear melting it. In addition the instrument is combined with a microscopy experiment. The setup consists of a solid state laser at a wavelength of 532nm as illumination source, two separated sending optics, a double-arm goniometer with two detection optics and an index match bath for the sample cell. The device is mounted on a vibration-free optical table. The main advantages of this setup are the use of counterpropagating illumination and a double-arm goniometer. To realize a counterpropagating illumination, the laser light is split by a beam splitter and injected into two different sending optics mounted on the optical table. The setup is equipped with individual illumination and detecting optics, for SLS and DLS, respectively. Separated illumination and detection optics allow to combine the use of different scattering experiments. In the sample cell both laser beams are collinear with their width and the detection optics optimized for

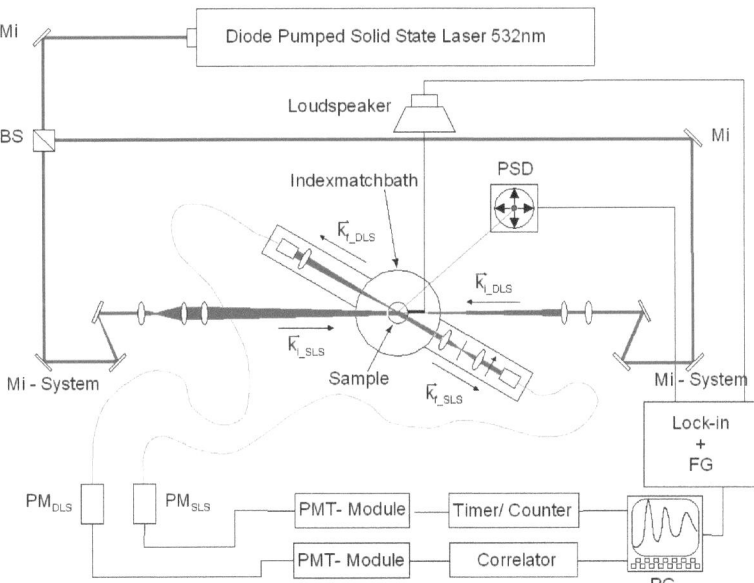

Figure 3.2: A sketch in top view of the setup. The basic construction of the multipurpose device in particular the double-arm goniometer for the simultaneous measurements of static, dynamic and elastic properties was developed by H. J. Schöpe [111, 112]. The figure includes following abbreviations: mirror (Mi), beam splitter (BS), photomultiplier (PM) and position-sensitive detector (PSD).

three experiments. The laser beam for the static light scattering (SLS) and the torsional resonance spectroscopy (TRS) enters from the left side ($\vec{k}_{i\_SLS}$) and the beam for the dynamic light scattering (DLS) enters from the right side ($\vec{k}_{i\_DLS}$). The scattered light $\vec{k}_{f\_SLS}$ and $\vec{k}_{f\_DLS}$ is detected by a photomultiplier.

The central bore of 20mm in diameter in the goniometer is for the flow through cell (outer radius 5mm, inner radius 4mm) made of quartz glass (refractive index $\nu = 1.458$) which is additionally surrounded by an index match bath also made of quartz glass (outer radius of 42.5mm, inner radius of 40mm). Index matching is necessary to avoid parasitic reflections at the inner sample cell and outer match bath surfaces consisting of quartz glass. It is achieved by silicon oil (refractive index $\nu = 1.453$ at $T = 20°C$).

The multi purpose instrument is exceptional in that also elastic properties of colloids can be investigated using TRS. Therefore, sample oscillations are achieved by coupling the cell to a commercial loud-speaker using an aluminum rod. The position-sensitive detector used for TRS can be adjusted to collect scattering signals stemming from a single Bragg reflection. For further analysis of the collected light, a standard lock-in technique is used (Stanford Research Systems, SR850 Sunnyvale USA). A resonance spectrum is recorded showing response amplitude and phase lag.

The precise control of the interaction in charge stabilized systems is important for a systematic analysis of the physical properties of colloidal solids. Therefore, utilization of a continuous conditioning technique is a prerequisite for each measurement. It is used to guarantee a fast and reproducible adjustment of interaction parameters by varying and controlling the particle number density and the counter ion concentration.

Figure 3.3: Photograph of the central double arm goniometer unit using a circular tracking system which comprises the detection for static and dynamic light scattering (SLS, DLS) and a position sensitive detector (PSD) for torsional resonance spectroscopy (TRS). In the middle of the goniometer the sample cell is index matched.

3.2.1 Static light scattering

For SLS experiments a broad parallel beam (ca. 4mm) is required to guarantee a good powder averaging. A sending optics system of three achromatic lenses (focus lengths: $f_1 = 20mm$, $f_2 = 100mm$ and $f_3 = 500mm$) is therefore used to obtain a slightly focused beam. For SLS detection two lenses ($f_1 = f_2 = 15mm$) and a small iris aperture is used to collect as much light as possible scattered at a well-defined scattering angle. A step wise increase of the detection angle is optimally suited for angle dependent structure identification. Statistical accuracy is considerably enhanced by rotating the sample about its vertical axis. The scattered light is focused on multimode fiber and transmitted to a photon-counting photomultiplier tube[1] (H7155 Hamamatsu Photonics, Deutschland GmbH). The photon-counting photomultiplier tube is connected to a National Instruments PCI-6030E multifunction data acquisition card through a NI-BNC-2110 connector block for the measurement data transfer to the PC. The communication between the PC and the connector block is programmed in a LabView environment.

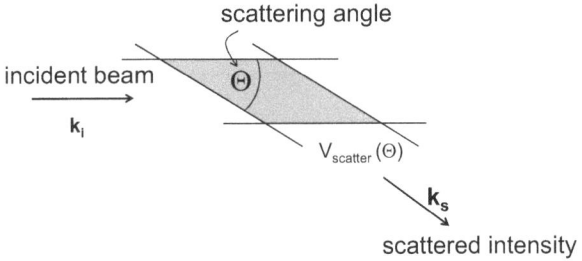

Figure 3.4: Angle dependence of the scattering volume with its lowest value at a scattering angle of $\Theta = 90°$.

For the goniometer of the light scattering setup an angle range between 10° and 170° is accessible. The scattered intensity $I_{\text{det}}(\Theta)$ is detected in dependence of the scattering angle Θ. First of all, the angle dependence of the scattering volume has to be corrected. The scattering volume has its lowest value at $\Theta = 90°$. Each deviation enlarges the volume by the factor $\sin(\Theta)^{-1}$ as demonstrated in Fig. 3.4. For eliminating this kind of angle dependence for the scattering volume, the detected intensity has to be multiplied by the same factor

$$I(\Theta) = I_{\text{det}}(\Theta) \cdot \sin(\Theta). \tag{3.1}$$

From the scattering angles, the scattering vector q is calculated:

$$|\vec{q}| = \left|\vec{k}_s - \vec{k}_i\right| = \frac{4\pi\nu_s}{\lambda} \sin\left(\frac{\Theta}{2}\right), \tag{3.2}$$

where λ denotes the laser wave length and ν_s the refractive index of the suspension's medium.

The q-values of the Bragg peaks with the corresponding Miller indices hkl are necessary to identify the lattice constant g. The next-nearest distance d_{hkl} of crystallographic planes for cubic systems is defined as

$$d_{\text{hkl}} = \frac{g}{\sqrt{h^2 + k^2 + l^2}}. \tag{3.3}$$

---

[1] The photomultiplier tube (PMT) module is basically comprised of a photomultiplier to convert light into electrical signals, a high voltage power supply circuit and a voltage divider circuit to distribute the optimum voltage to each dynode, all assembled into a single compact case. It has a spectral response between 300 and 650nm in wavelength. This photon counting method is superior to analog signal measurements in terms of stability, detection efficiency and signal-to-noise ratio [117].

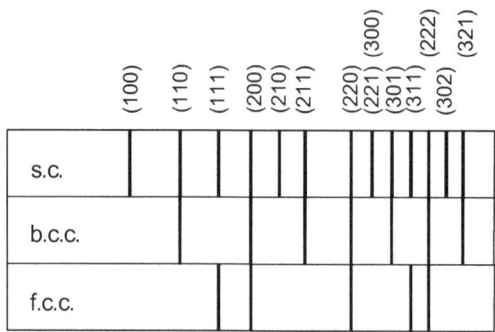

Figure 3.5: A scheme of Bragg reflections for different cubic crystal structures: For fcc, the rule is that h, k, and l must be all even or all odd. For bcc, $h+k+l$ must be even. For a sc crystal, all peaks are 'allowed' (figure taken from [88]).

With the Bragg condition $2\pi = |\mathbf{q}|\, d_{hkl}$ this equation yields

$$|\mathbf{q}| = \frac{2\pi\sqrt{h^2 + k^2 + l^2}}{g} \qquad (3.4)$$

for the absolute value of the reciprocal lattice vector, which corresponds to the Bragg reflection. According to the values of the Bragg peaks and the right selection rule in dependence of the assumed cubic structure, the lattice constant g, and the particle number density n can be determined. There are actually three different kinds of cubic crystals: simple cubic (sc) in which the same pattern (or basis) of atoms is repeated in each cube, face-centered cubic (fcc) in which the same pattern is repeated on the corners and the center of each face of the cube, and body-centered cubic (bcc) in which the same pattern is repeated at the corner and the center of the cube. For a sc crystal, all peaks are 'allowed'. But fcc and bcc crystals have selection rules to determine which peaks are nonzero. For fcc, the rule is that h, k, and l must be all even or all odd. For bcc, the rule is that $h+k+l$ must be even. A scheme of Bragg reflections for different cubic crystal structures are shown in Fig. 3.5.

The particle number density n for the three primitive cells in cubic structures has following relation to the number of particles per cell:

$$\begin{aligned} n &= \frac{1}{g^3} \quad \text{(sc: 1 particle per unit cell)} \\ n &= \frac{2}{g^3} \quad \text{(bcc: 2 particles per unit cell)} \\ n &= \frac{4}{g^3} \quad \text{(fcc: 4 particles per unit cell)}. \end{aligned} \qquad (3.5)$$

Fig. 3.6 shows exemplarily an intensity distribution $I(q)$, which provides information on the structure of the sample. The Bragg peaks can be indexed by assuming a bcc structure with the lattice constant of $g = 1.05\,\mu m$ and the particle number density $n = 1.7\,\mu m^{-3}$. It shows a typical diffraction pattern of a polycrystalline sample, where the crystallites are randomly oriented. The diffraction produced by such a sample is what one would obtain by combining the diffraction patterns of all possible orientations of a single crystal.

The diffraction pattern of a polycrystalline sample depends on the number of crystallites and their size in the scattering volume. With increasing number of crystallites, closed diffraction rings (Debye-Scherrer rings) appear around the beam axis rather than the discrete Laue spots which are typical

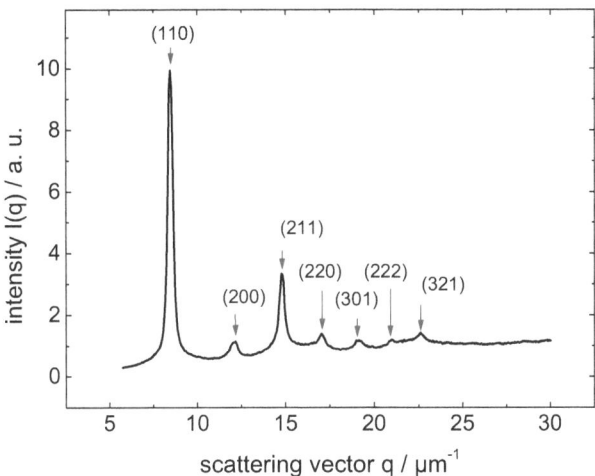

Figure 3.6: An example of a diffraction pattern for a colloidal suspension of Poly-n-Butylacrylamide copolymer particles (PnBAPS) with a diameter of 122nm. The peaks are indexed under the assumption of a bcc structure. The obtained lattice constant is $g = 1.05\mu m$ and the particle number density $n = 1.7\mu m^{-3}$.

of single crystal diffraction. The width of the Debye-Scherrer rings depends on the number of atomic planes per crystallite which are involved in the scattering process. The known number of atomic planes can be simply related to the crystal size. In the case of cubic systems this assumption leads to [118–122]:

$$\Delta q = \frac{2\pi K}{L}. \quad (3.6)$$

L is the average length of a cubic shaped crystallite (or coherent length of the crystal lattice), Δq the full width at half height of a Debye-Scherrer ring and K the Scherrer constant which is of the order of unity for cubic crystals. The constant K depends on the shape of the particles and their structure. For spherical particles and cubic structure, Scherrer arrived at a value of $K = 0.89$ [118] and Patterson obtained $K = 1.11$ [121], so that a value of $K = 1$ is used as a compromise [122]. Note that the use of the Scherrer analysis by Eq. (3.6) to determine crystallite size generally gives a lower bound on the true crystallite size. Lattice distortions or instrumental contributions can cause additional peak broadening.

3.2.2 Dynamic light scattering

The Static Light Scattering technique measures the ensemble-averaged intensity of scattered light over a large time scale where the motion of each particle is averaged out and not taken into account. This motion arises from the fact that the colloidal particles are undergoing rapid thermal motions due to collisions between the colloidal particles and the molecules of the solvent. These movements are called Brownian motion and they cause short term fluctuations in the intensity of the scattered light. The Dynamic Light Scattering (DLS) technique measures the time-dependent fluctuations in the intensity of scattered light. The intensity of the collected light depends on the interference from light scattered off of different particles. As these particles move relative to each other, the interference condition changes. It is the statistics of the changing interference condition that is related to the motion of the particles.

The experimental performance of DLS measurements differs from static light scattering experiments. While in the latter case the sample must be illuminated with a broad beam to obtain a good average over the sample, DLS requires a narrow incident beam. For DLS illumination two transmission lenses ($f = 500mm$ and $f = 40mm$) are used (see Fig. 3.2), in a manner that a beam waist radius of about $100\mu m$ in the middle of the sample is achieved and a single speckle[2] can be observed with monomode fiber optical detection [123, 124]. One detection lens ($f = 250mm$) and a fiber coupler with a single mode fiber is therefore used. Detection is performed using a photomultiplier module (in the same way as for SLS) for single photon counting. The resulting output signal is fed into a digital correlator (ALV 7004, ALV-GmbH, Langen), which determines the autocorrelation function of the time-dependent scattering signal corresponding to the intensity fluctuation of the speckle.

The information of the time-dependent spatial fluctuations can be analyzed by constructing the field autocorrelation function which is the time average (t_m) of the correlation between a field measured at time t and the same field measured at time $t + \tau$:

$$G^{(1)}(\mathbf{q}, \tau) = \langle \mathbf{E}(\mathbf{q}, t)\mathbf{E}(\mathbf{q}, t+\tau) \rangle = \lim_{t_m \to \infty} \frac{1}{t_m} \int_0^{t_m} \mathbf{E}(\mathbf{q}, t)\mathbf{E}(\mathbf{q}, t+\tau) dt \qquad (3.7)$$

While the field autocorrelation function is not directly measurable, the intensity autocorrelation function is experimentally accessible:

$$G^{(2)}(\mathbf{q}, \tau) = \langle I(\mathbf{q}, t)I(\mathbf{q}, t+\tau) \rangle = \lim_{t_m \to \infty} \frac{1}{t_m} \int_0^{t_m} I(\mathbf{q}, t)I(\mathbf{q}, t+\tau) dt \qquad (3.8)$$

The field and intensity autocorrelation function are normalized by their values at zero τ:

$$g^{(1,2)}(\mathbf{q}, \tau) = \frac{G^{(1,2)}(\mathbf{q}, \tau)}{G^{(1,2)}(\mathbf{q}, 0)} \qquad (3.9)$$

The assumption of a Gaussian form for the field autocorrelation function enables the determination of the intensity autocorrelation function [125]. For an ergodic system the intensity correlation function $g^{(2)}(q, \tau)$ and the field correlation function $g^{(1)}(q, \tau)$ are connected via the Siegert relation:

$$g^{(2)}(q, \tau) = 1 + \beta |g^{(1)}(q, \tau)|^2, \qquad (3.10)$$

where β is the intercept of the correlation function with the maximal value 1 and τ is the correlation time. It is reduced by alignment errors or detector nonlinearities. The functional form of $g^{(1)}(q, \tau)$ is determined by the kind of dynamics present in the suspension. In the simplest case of noninteracting identical particles all motion is due to diffusion. This case is encountered in highly dilute colloidal systems and is described by the Stokes Einstein self-diffusion coefficient D_0 [39, 123]:

$$D_0 = \frac{k_B T}{6\pi \eta a_H}, \qquad (3.11)$$

where a_H is the hydrodynamical radius and η the viscosity of the solvent. In this case the field autocorrelation function reads:

$$g^{(1)}(q, \tau) = \exp(-q^2 D_0 \tau). \qquad (3.12)$$

The diffusion coefficient D_0 is calculated by fitting the correlation curve $g^{(1)}(q, \tau)$ with D_0 showing proportional behavior to the correlation time of the exponential decay. The hydrodynamical radius a_H is then calculated using the Stokes Einstein equation (3.11) and is an important parameter to characterize the particle size of a colloidal suspension.

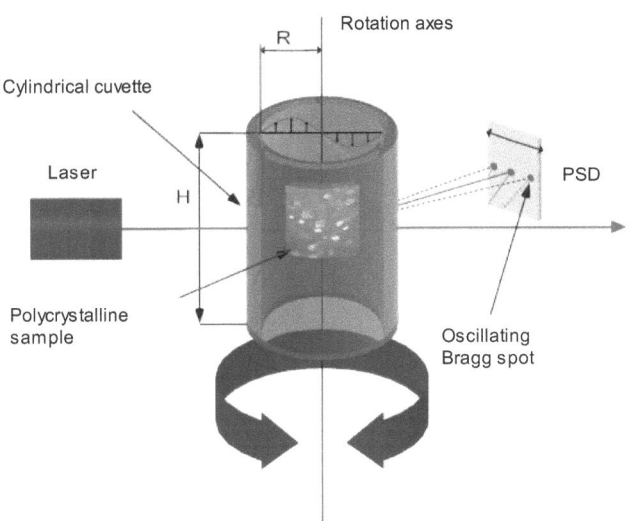

Figure 3.7: A scetch of the torsional resonance spectroscopy experiment: a cylindrical cuvette with a height H and a radius R is excited by shear induced oscillations. On the colloidal level this corresponds to harmonic lattice vibrations visible as shifts of the Bragg peak (figure taken from [6]).

3.2.3 Torsional resonance spectroscopy

Torsional resonance spectroscopy (TRS) enables the analysis of elastic properties. Measurements of the shear modulus in polycrystalline colloidal systems were introduced by Joanny and Pieranski [95, 96, 109] for the first time. Further development of the detection scheme was achieved by Palberg et al. [126].

In TRS experiments the sample cell is put into low frequency oscillations about its vertical axis, which excite the solid in the cell to resonant vibrations, if the eigen frequency for the given geometry is met (see figure 3.7). As the shear moduli of the sample are small, standing waves can be excited with wavelengths comparable to the container dimensions. Torsion of only a tenth of a degree and frequencies in the range of 0.5-25Hz lead to periodically shear stressed vibrations in the sample induced by a loudspeaker. A position sensitive detector (PSD) is placed to detect a single Bragg reflection of an individual crystallite. Due to the excited shear waves, the lattice constant of the oscillating crystal is altered periodically. This corresponds to harmonic lattice vibrations resulting to periodic shifts of the Bragg spots.

In the experiment, a frequency ramp is continuously applied and the amplitude and the phase of a single oscillating Bragg spot is detected by a PSD. For further analysis of the collected light, a standard lock-in technique is applied (Stanford Research Systems, SR850 Sunnyvale USA). Lock-in amplifiers are used to measure the amplitude and phase of signals embedded in noise. They achieve this by acting as a narrow bandpass filter which removes much of the unwanted noise while allowing the signal to pass through which is to be measured. The frequency of the signal to be measured and hence the passband region of the filter is set by a reference signal, which has to be supplied

---

[2] Objects with rough surfaces on the scale of wavelengths of visible light (or colloidal suspensions) illuminated by light from a highly coherent laser are readily observed to acquire a granular appearance. Under illumination by coherent light, the wave reflected from such a surface consists of contributions from many independent scattering areas. Interference of these dephased but coherent waves results in the granular pattern also known as speckle.

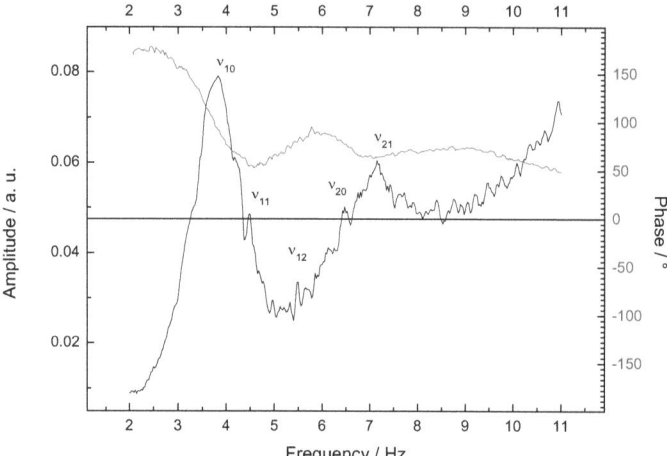

Figure 3.8: Characteristic resonance spectrum of a colloidal suspension consisting of polystyrene spheres with 122nm in diameter at the particle concentration of $n = 48.1 \mu m^{-3}$ crystallized in a bcc structure. The spectrum shows the characteristic resonance frequencies and the phase with the corresponding shear modulus of $G = 0.5 Pa$.

to the lock-in amplifier along with the unknown signal. The lock-in amplifier SR850 has its own digitally synthesized reference source. Because the internal reference source is synthesized from the same digital signal that is used to multiply the input, there is virtually no reference phase noise.

Fig. 3.8 shows a characteristic resonance spectrum of a crystallized colloidal suspension consisting of polystyrene spheres with 122nm in diameter at the particle concentration of $n = 48.1 \mu m^{-3}$ with bcc structure. In resonance, the phase lag between a signal used to excite the sample to torsional oscillations (vibrations) and the response of the sample itself vanishes and the amplitude is increased. Two evaluation methods exist to determine the resonance frequency. First, a Lorentzian curve is fitted to the amplitude at the resonance frequency. Alternatively, an arctan function is fitted to the decrease of the phase at the point of inflection (in the range of the maximum amplitude) [111]. The spectrum in Fig. 3.8 shows the characteristic resonance frequencies and the phase with the corresponding shear modulus of $G = 0.5 Pa$.

The resonance frequency ν_{10} can be identified and related to the shear modulus G following Eq. (2.74). In principle, also resonances of higher order could be taken for the calculation of the shear modulus. The disadvantage in this case is that they are smeared and broadened by the presence of secondary order resonances and the determination of their exact position is more difficult.

3.3 Time-resolved microscopy

Microscopical investigations of the crystal growth velocities can be performed at low particle number densities where heterogeneous nucleation dominates the solidification process. Under these conditions single crystals nucleate heterogeneously at the cell wall after cessation of shear (see figure 3.9). Crystallization under oscillatory mechanical shear leads to the formation of twinned oriented wall crystals [103]. The direction of easiest shear within the $\langle 111 \rangle$ planes is oriented parallel to the formerly applied shear flow direction. After cessation of shear flow, a planar front of twinned bcc crystals propagates linearly with their densest packed planes perpendicular to the cell wall in $\langle 110 \rangle$ direction [76, 127].

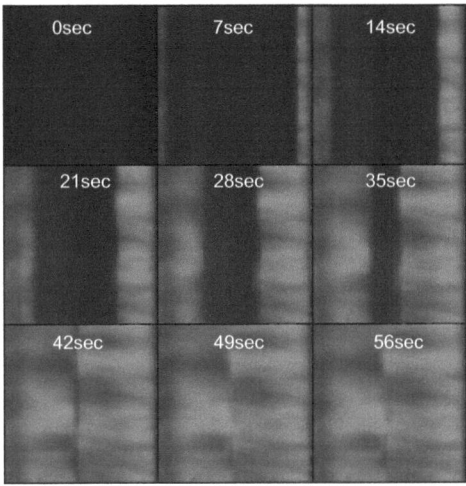

Figure 3.9: Microscopic images of the solidification process of a colloidal silica suspension with particles of 86nm diameter at low particle concentration $n = 19.0 \mu m^{-3}$. The images were taken by polarization microscopy in side view of the sample cell. The colored region is the wall crystal growing linearly from both cell walls into the center of the sample cell. The image size is $1.00 \times 1.16 mm^2$. The corresponding growth velocity is determined to be $10.7 \mu m/s$.

Solidification is monitored by polarization Bragg-microscopy [128–130] in a flat flow through cell made of quartz glass with a wall-to-wall distance of 1mm. The cell is mounted on the stage of an inverted microscope (Leica DMI3000 B, Leica, Wetzlar, Germany) equipped with a low resolution objective (5 × 0.09, Leica, Wetzlar, Germany). Images are recorded by a charged-coupled device (CCD) camera and stored in a PC for image analysis.

Fig. 3.9 shows microscopic images of the solidification process of a colloidal silica suspension at low particle concentration $n = 19.0 \mu m^{-3}$ with 86nm particle diameter taken by polarization microscopy in side view of the sample cell. The colored region is the wall crystal growing linearly from both cell walls into the center of the sample cell. The image size is of $1.00 \times 1.16 mm^2$. The corresponding growth velocity is determined to be $10.7 \mu m/s$.

At high particle number densities or increasing interaction between the particles, homogeneous nucleation dominates the solidification process. In this case randomly oriented crystals appear. A typical image of a polycrystalline sample solidified from an undercooled melt at $n = 32.0 \mu m^{-3}$ is shown in Fig. 3.10. The crystals appear as faceted and irregularly shaped polyeders of different colors. The facets result from crystal intersections occurring during growth. The color differences originates from different crystal orientations [128, 130].

3.4 Ultra Small Angle X-ray Scattering

In addition to light scattering experiments also Ultra Small Angle X-ray Scattering (USAXS) at the beamline BW4 at the Hamburger Synchrotronstrahlungslabor HASYLAB (DESY) was used to analyze the short-range order and the nucleation behavior in colloidal suspensions. USAXS makes use of the penetration of X-rays through materials, either in solid or liquid state. This feature renders USAXS an invasive but not destructive method. While wide-angle X-ray scattering probes electron density fluctuations on the length scales of interatomic distances, typically scattering angles of $2\theta < 1°$ are considered as the USAXS regime. For these scattering angles x-rays probe long-range

Figure 3.10: Microscopic image of a solidified colloidal silica suspension with particles of 84nm diameter at a particle number density $n = 26.0 \mu m^{-3}$ taken by polarization microscopy in top view of the sample cell. The image size is $1.55 \times 1.16 mm^2$

fluctuations of the mean electron density over many atoms, thereby giving access to correlation length scales ranging to several micrometers.

3.4.1 Beamline BW4 at HASYLAB (DESY)

The beamline BW4 at HASYLAB is designed as a transmission (Ultra) Small-Angle X-ray Scattering ((U)SAXS) instrument with a high flux material research setup [131]. The BW4 is a X-ray wiggler ($N = 19$ periods, $K = 13.2$)[3] beamline with instrumentation to measure (Ultra) Small Angle X-ray Scattering. A well collimated monochromatic X-ray beam is guided to a sample and the intensity of the radiation which is scattered from this sample under very small angles relative to this incident beam recorded as a function of the scattering direction. The beamline starts at the location of the wiggler and has a total length of 56 m (see Fig. 3.11). The tilt absorber is the only element in the beamline capable to absorb the total power of the white wiggler beam. The horizontally focusing optical element (H-focussing mirror) is a water cooled platinum coated silicon mirror with fixed cylindrical shape. The BW4 has a fixed-exit double-crystal-monochromator with Si(111) crystals leading to a standard wavelength of $\lambda = 0.138nm$. The vertically focusing optical element (V-focussing mirror) is a plane mirror installed in a mirror bending device to realize a curved mirror surface. The guard slit system (Slit 1 and 2) consists each of four silicon slit blades driven by a piezo motor with an accuracy of $0.1 \mu m$. This slit system produces a beam of $400 \times 400 \mu m^2$ with a Gaussian beam profile. The blades have a special geometry and surface treatment to reduce scattering from the slits [132].

Typical for transmission USAXS, is the use of a beam stop obscuring the direct beam. A photodiode in the center of the beam stop is used to monitor the primary beam intensity passed through the sample (see Fig. 3.13). The resolution is limited by the minimum size of the beam stop resulting in the maximum detectable correlation distance of $d_{\max} = 650nm$. The beam stop prevents the

---

[3] A wiggler is a periodic structure of dipole magnets with alternating poles and N magnetic field periods. It produces a periodic transverse magnetic field causing the electrons (or positrons) in the storage ring to follow a sinusoidal path with a wavelength λ. A wiggler is characterized by the dimensionless parameter $K = eB\lambda/2\pi m_e c_0$, where e is the electron charge, B the magnetic field, m_e the electron mass and c_0 the velocity of light. If $K > 1$ the amplitude is large and the radiation contributions from each field period sum up independently, leading to a broad energy spectrum. If $K < 1$ the oscillation amplitude is smaller and the radiation displays interference patterns leading to narrow energy bands. In the latter case, the device is called an undulator.

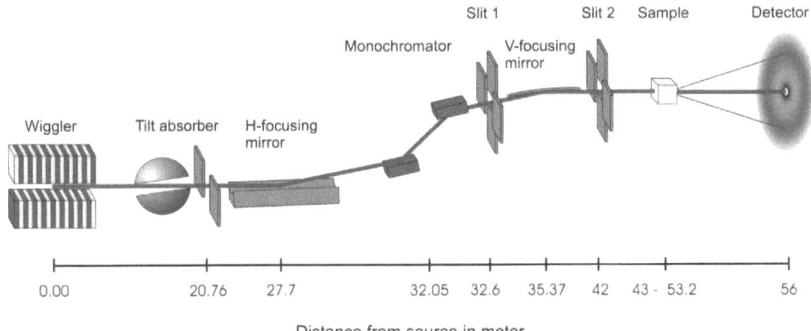

Figure 3.11: The beamline BW4 at HASYLAB (DESY) [132]. The beamline BW4 starts at the location of the wiggler and has a total length of 56 m. The tilt absorber is the only element in the beamline capable to absorb the total power of the white wiggler beam. The horizontally focusing optical element (H-focussing mirror) is a water cooled platinum coated silicon mirror with fixed cylindrical shape. The BW4 has a fixed-exit double-crystal-monochromator with Si(111) crystals. The vertically focusing optical element (V-focussing mirror) is a plane mirror installed in a mirror bending device to realize a curved mirror surface. The guard slit system (Slit 1 and 2) consists each of four silicon slit blades driven by a piezo motor with an accuracy of $0.1\mu m$. The blades have a special geometry and surface treatment to reduce scattering from the slits.

detector from being damaged by the direct beam. For monitoring the intensity of the incident primary beam an ionization chamber is provided. Two types of detectors are available for detecting the scattered intensity. The first is a commercial CCD-detector (MARCCD165) [133] with an active diameter of 165mm and a resolution of 2048×2048 (pixel size $79.1 \times 79.1 \mu m^2$) which operates with a read-out time of 3.5 sec. In addition, a PILATUS100K [134] detector with an active area of $84 \times 34mm$ and a resolution of 487×195 (pixel size $172 \times 172 \mu m^2$) with a read-out time of $5 \cdot 10^{-3} sec$ was utilized. PILATUS detectors have several advantages in comparison to CDD and image plate detectors including a high frame rate up to 100Hz, no read-out noise, excellent signal-to-noise ratio and high detected quantum efficiency. These properties ensure a good image quality even at short exposure times.

A key parameter characterizing SAXS measurements is the resolution in terms of the minimum accessible scattering angle θ_{min}. By Bragg's equation it is related with the maximum correlation distance d_{max} accessible in real space:

$$d_{max} = \frac{\lambda}{2\sin\theta_{min}} \approx \frac{\lambda}{2\theta_{min}}, \qquad (3.13)$$

where λ denotes the used wavelength of x-rays. The minimum scattering angle of the BW4 setup is on the order of $0.01°$.

Since particle distances below 360nm are expected in the analyzed colloidal system, the largest possible sample-to-detector distance was chosen and calibrated with a collagen sample to be 13.3m. With these parameters the accessible range of scattering vectors is $0.008nm^{-1} < q < 0.280nm^{-1}$.

Samples were investigated using a flow through cell shown in Fig. 3.12 (d). This sample cell was adjusted to the geometry of the sample chamber at BW4 and is connected on both sides for the computer controlled preparation circuit as introduced in section 3.1. Kapton® of $25\mu m$ of thickness

Figure 3.12: Beamline BW4 at Hasylab (DESY). Figure (a) shows the vacuum tube of the beamline BW4 which guides the scattered radiation to the detector with its largest possible distance $L_{SD} = 13m$ between the sample chamber (b) and the Pilatus detector (c). L_{SD} was calibrated with a collagen sample. Figure (d) shows the flow through cell used for USAXS measurements. At both sides the connections for the computer controlled preparation circuit are seen. A Kapton® film of thickness $25\mu m$ and a diameter of 4mm was used as window assuring good transmission at $\lambda = 0.138 nm$. The sample volume is about $150 mm^3$ and the wall-to-wall distance is 3mm.

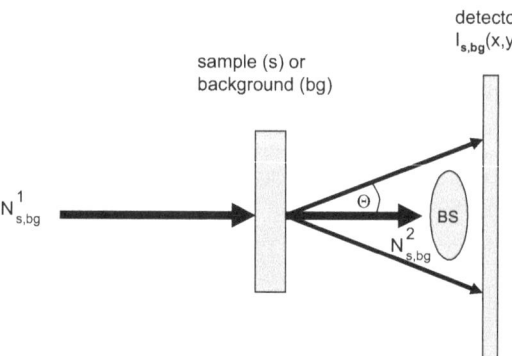

Figure 3.13: For monitoring the intensity of the incident beam an ionization chamber in front of the sample and for the transmitted intensity a photo diode on the beam stop are provided. The 2D scattering signal is optionally detected by two detectors: MARCCD165 [133] or PILATUS [134].

was chosen as window material. Kapton® has a good transmission for x-rays and is easy to handle in contrast to Beryllium. Beryllium is often used as window material for x-ray measurement facilities due to the highest degree of transmission for X-rays. The disadvantage is the toxic influence that requires intensified safety measures in the laboratory and at the beamline. The diameter of the window is 4mm and the wall-to-wall distance is 3mm leading to a probe volume of 150 mm^3.

By a computer control the electromagnetic valves in the preparation circuit stop the flow instantaneously indicating the start of a crystallization experiment. A LabVIEW program was developed to control the complete preparation circuit during the USAXS measurements, where the setup is inaccessible inside the experimental hutch. For each set of interaction parameters a sequence of diffraction patterns was taken with an acquisition time of 0.1 to 5sec depending on the scattering intensity and the used detector.

3.4.2 Evaluation of diffraction measurements

For the pre-evaluation of scattering data, the program FIT2D is used, which is offered by the European Synchrotron Radiation Facility (ESRF) as free ware [135]. First of all, every scattering pattern has to be divided by the actual incident flux measured by the ionization chamber and by the exposure time t. Further evaluation of the diffraction patterns includes a background and transmission correction of the 2D scattering intensities which is a standard procedure described in detail in [136]. For the background correction, the scattering intensity $I_{bg}(x,y)$ of the solvent and of the window material without the colloidal suspension is measured and has to be subtracted from the scattering profiles measured with the colloidal sample. The transmission correction considers the sample

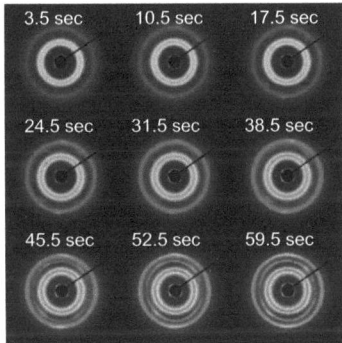 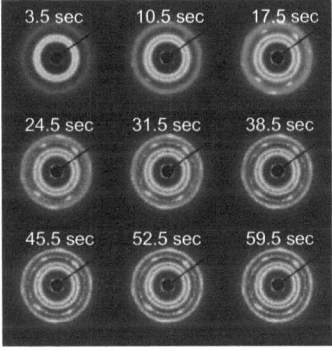

Figure 3.14: The left picture shows a sequence of nine diffraction patterns of the colloidal silica suspensions with particle diameter of 84nm taken within one minute at a particle number density $n = 113 \mu m^{-3}$ and sodium hydroxide concentration $c_{NaOH} = 1.4 \times 10^{-3} mol/l$ at strongest particle interaction where homogeneous nucleation dominates the crystallization behavior. The right picture shows a sequence of diffraction patterns at identical particle number density and identical accumulation time but at a much lower NaOH concentration $c_{NaOH} = 0.4 \times 10^{-3} mol/l$. From the beginning of crystallization a weakly developed four-fold point pattern is visible which can be ascribed to an oriented wall based single crystal originated from heterogeneous nucleation [88, 104] which is more pronounced in the state of lower interaction strength.

thickness and the fact that X-rays are always absorbed when they pass through matter. As a result, the total transmitted intensity I_t measured after passing the sample is only a fraction I_t/I_0 of the incident intensity I_0. In combination with the absorption law, following absorption and background correction has been applied:

$$I_{\text{corr}}(x,y) = I_s(x,y)\exp\left(\frac{\mu l}{\cos 2\theta}\right) - I_{\text{bg}}(x,y), \tag{3.14}$$

where $I_s(x,y)$ and $I_{\text{bg}}(x,y)$ are the measured 2-dim intensities with and without the colloidal sample. Here μ is the linear absorption coefficient (as a function of the X-ray wavelength and the chemical composition of the sample) and l is the sample thickness. The intensity is increasingly dampened with increasing scattering angle. Simplification is possible in the case of small scattering angles ($2\theta \to 0$ and thus $\cos 2\theta \to 1$) obtaining the linear absorption factor $\exp(-\mu l)$. The experimental determination of the absorption factor is based on measurements of photon counts before (N^1) and behind (N^2) the sample (see figure 3.13). In this case, $\exp(-\mu l) \approx \left(\frac{N_s^2}{N_s^1}\right)/\left(\frac{N_{\text{bg}}^2}{N_{\text{bg}}^1}\right)$ is approximately valid leading to the following expression for the transmission and background correction:

$$I_{\text{corr}}(x,y) = I_s(x,y)\left(\frac{N_s^2 N_{\text{bg}}^1}{N_s^1 N_{\text{bg}}^2}\right) - I_{\text{bg}}(x,y), \tag{3.15}$$

The next evaluation step includes an azimuthal or the so called 2θ integration of the corrected intensity of each single 2D diffraction pattern. Azimuthal averaging means that for a chosen distance from the center of the pattern all the pixels on a circular ring are picked. Their intensity is summed and divided by the number of the valid pixels obtaining a scattering pattern that is proportional to the scattering intensity in absolute units.

Time resolved diffraction patterns of the shear molten and re-crystallizing colloidal silica suspension were systematically measured for several interaction strengths. Two examples of these 2-dim sequences are shown in Fig. 3.14 for a silica suspension (84nm particle diameter) at a particle number density $n = 113\mu m^{-3}$. The exposure time and the read-out time of the detector are both 3.5s in this case. The left figure represents the system at maximum interaction ($c_{\text{NaOH}} = 1.4 \cdot 10^{-3} mol/l$) where homogeneous nucleation dominates the crystallization process. The first diffraction pattern shows the metastable melt state. After the third diffraction pattern, the Debye-Scherrer rings become more narrow and additional rings appear indicating the transition to a polycrystalline bcc state. In addition, from the beginning of crystallization a weakly developed four-fold point pattern is visible which can be ascribed to an oriented wall based single crystal originated from heterogeneous nucleation [88, 104]. Heterogeneous nucleation is more pronounced in the state of lower interaction strength. This effect is shown in the right part of Fig. 3.14 where a small amount of added NaOH ($c_{\text{NaOH}} = 0.4 \times 10^{-3} mol/l$) is present. Here the appearance of heterogeneous nucleation has a larger influence on the scattering profiles.

The evaluation of the measurement data takes this effect into account by applying masks to isolate the scattering part deriving from the homogeneously nucleated polycrystalline material of the illuminated sample area. This evaluation procedure is demonstrated in Fig. 3.15. Fig. (a) shows a diffraction pattern including the raw data and Fig. 3.15 (b) demonstrates the same image including a background and transmission correction. Scattering signals are considered by an applied mask (Fig. 3.15 (c)) which excludes the contributions of scattering signals stemming from heterogeneously nucleated wall crystals. Note that in this special case as shown in Fig. 3.15 (c), the contributions of heterogeneous nucleation are excluded for the first Debye-Scherrer ring only and not necessarily for those of higher order. Nevertheless, this procedure is sufficient for the determination of nucleation parameters which are extracted from the first scattering peak only. A detailed description for this approach is given in sec. 4.4.

The diffraction patterns in Fig. 3.15 show one of the main advantages of USAXS measurements of colloidal systems, namely the discrimination of the scattering information resulting from polycrystalline material or oriented wall crystals. Wette et al. analyzed the competitive behavior between heterogeneous and homogeneous nucleation near a flat wall [137]. They observed that heterogeneous nucleation at the container walls is delayed in comparison to the homogeneous bulk nucleation and

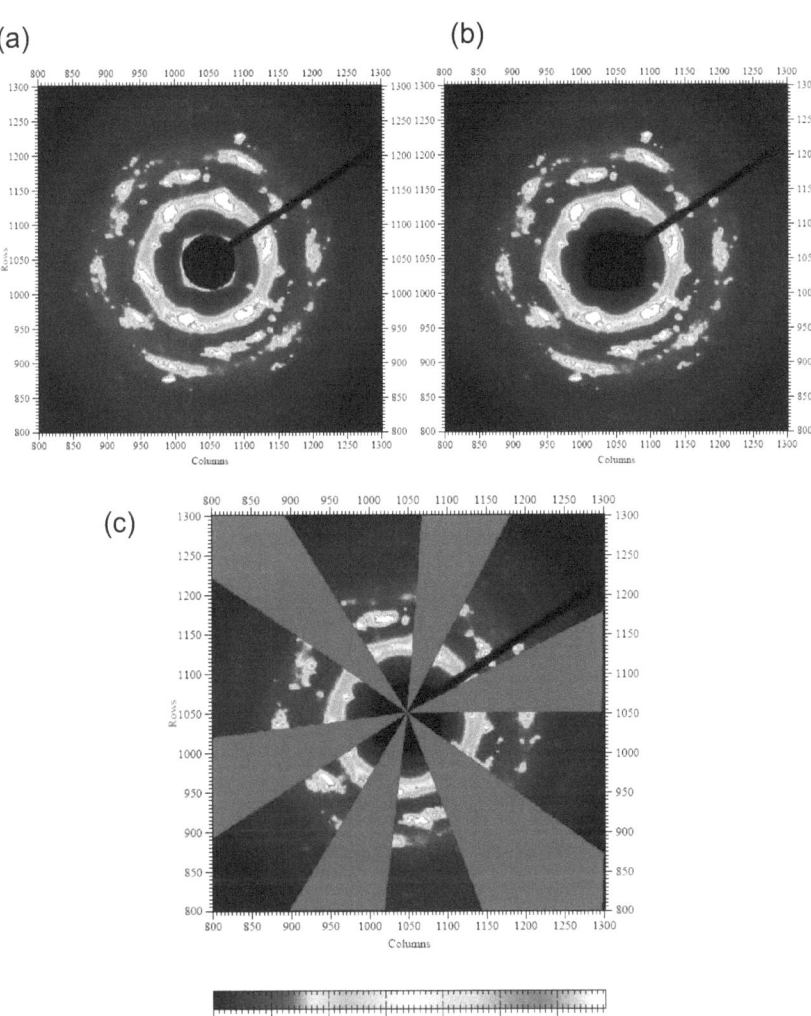

Figure 3.15: Diffraction pattern including the raw data (a), background and transmission corrected data (b) and the same with an applied mask (c) which excludes the contributions of scattering signals stemming from heterogeneously nucleated wall crystals. This evaluation method allows the discrimination of the scattering information resulting from polycrystalline material or oriented wall crystals.

the nucleation rate density appears surprisingly slightly smaller demonstrating the complexity of the observed crystallization process.

For the determination of the particle form factor $P(q)$, additionally the intensity at each analyzed particle number density with high amounts of NaOH added was recorded and evaluated in the same way as the sample scattering profiles. At high NaOH concentrations the electrostatic repulsion is screened due to the excess of counter ions. Then interactions are suppressed representing $S(q) = 1$, hence the particle form factor $P(q)$ is directly accessible. The background and transmission corrected

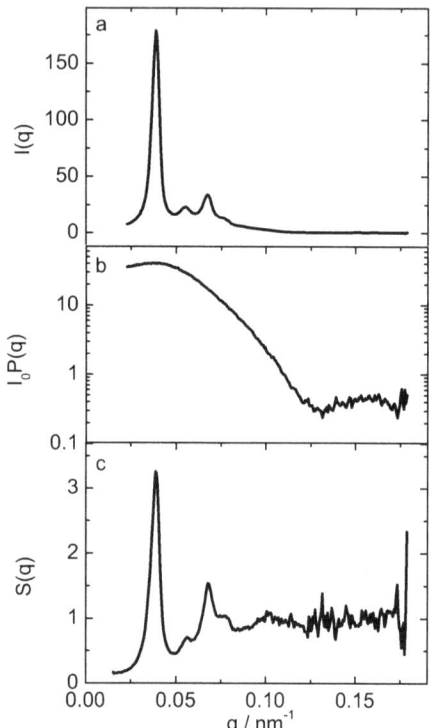

Figure 3.16: Experimentally determined intensity distribution of a silica suspension consisting of particles of 77nm diameter at the particle number density $n = 80.0 \mu m^{-3}$. The background and transmission corrected intensity $I(q)$ has to be divided by the particle form factor $P(q)$ distribution (b) to obtain the structure factor $S(q)$ (c).

intensity $I(q)$:

$$I(q) = I_0 \, P(q) \, S(q) \tag{3.16}$$

has to be divided by the particle form factor $P(q)$ distribution to obtain the structure factor $S(q)$. This evaluation procedure is shown in Fig. 3.16 where the averaged intensity distribution $I(q)$ (a) and the particle form factor $P(q)$ (b) of a silica suspension consisting of particles of 77nm diameter at the particle number density $n = 80.0 \mu m^{-3}$ are experimentally determined in two independent experiments. The evaluated structure factor $S(q)$ is shown in Fig. 3.16 (c).

Note that the structure factor $S(q)$ shows a broad crystalline (110) peak which reveals an average crystallite size of $L = 800 nm$. Compared to atomic systems the size seems to be high enough not to cause peak broadening due to a small coherence length for interference of the scattered light. However, it should be noted that charged colloidal systems are not dense packed. The scattering information which is shown in Fig. 3.16 corresponds to a silica suspension with a particle number density of $n = 80.0 \mu m^{-3}$ and a next nearest neighbor distance of $d_{NN} = 230 nm$. If one scales the crystallite size with the next nearest neighbor distance, a number of coherent layers yields in between 3-4. In addition, also the fluid carrier medium where the crystallites are embedded leads to a peak broadening.

Chapter 4
Results and discussion

Metals are non-transparent systems consisting of atoms, which show fast relaxation dynamics in the liquid state. Therefore, it is extremely difficult to directly observe e.g. solid-liquid interfaces, nucleation of crystalline phases and instabilities of a solidification front. In contrast, colloidal suspensions are transparent and the dynamics of their particles is much more sluggish than the relaxation of atoms in metals. Detailed investigations of phase transitions in colloids [138] are possible even in real-space [69]. Therefore, colloidal systems with tunable interactions are often discussed as model systems for atomic materials [7]. The scientific target of the present thesis is to analyze the potential of colloids as model systems with respect to the solidification of metals.

This chapter begins with a detailed characterization of the investigated colloidal system consisting of charged silica spheres dispersed in pure water. In particular, its unique property of tunable interaction is emphasized, since it is one of the main advantages in colloids with respect to atomic or molecular systems, where the potential is fixed due to the properties of the system. The complete characterization including detailed investigations on the phase diagram, crystal growth and elastic behavior is followed by studies on short-range order and nucleation behavior in dependence of the variable interaction potential. To proof its model character for metals the results as the short-range order in the metastable colloidal melt are compared to those obtained for an undercooled Nickel melt [3] revealing an icosahedral short-range order. The investigations on the short-range order were focussed on the metastable state at large undercoolings. Additional studies have been performed in the equilibrium state by analyzing the short-range order near the solid-fluid phase boundary.

Further, investigations on the nucleation behavior have been performed. The results were analyzed within the framework of classical nucleation theory following a formalism with adaption to colloidal systems [4,5]. This procedure gives access to an important property for solid-fluid transitions, namely the interfacial energy in dependence of the degree of metastability. These results can be compared to metals by performing a Turnbull plot to discuss the model character of colloids concerning nucleation phenomena [6].

4.1 Interaction control in charged sphere silica systems

The development of simple and versatile methods for the preparation of nanoparticles of a well defined size or shape is an important and challenging task for the design of particles with novel physical properties. The manifoldness of colloidal systems in terms of size and shape of the particles requires flexible and variable synthesis methods [139].

For detection of the colloidal melt structure and nucleation behavior, particles of high scattering contrast for X-rays had to be chosen. For instance, polystyrene spheres or PMMA (polymethyl methacrylate) particles are not suitable due to the low scattering contrast between water and polystyrene. A possibility to increase the contrast is to use methanol or methanol/water mixtures as a solvent system. The methanol does not affect the refractive index but the interaction between the particles is reduced to a large extent due to the low lying permittivity of methanol. Due to the high electron density contrast in aqueous dispersions, colloidal silica systems make detailed

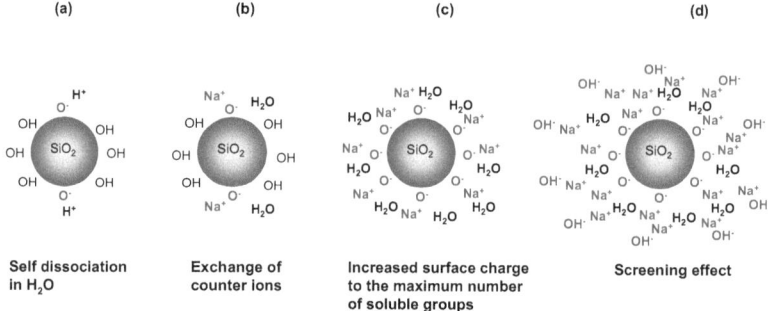

Figure 4.1: Control of silanol surface groups on silica particles: (a) Self dissociation of the silanol groups in aqueous dispersion; (b) Exchange of counter ions of H^+ to Na^+ by addition of NaOH; (c) Charging up reaction to the maximum number of soluble surface groups corresponding to the state of maximum interaction; (d) Screening effect caused by further addition of NaOH after the equivalent point is crossed.

ultra-small-angle X-ray scattering studies possible even at low volume fractions which characterize charge stabilized colloidal systems. First intensive experiments in this field were made by Konishi et al. [140–142]. For this reason charge stabilized colloidal silica systems synthesized by means of a modified Stöber synthesis [143] were chosen.

This kind of synthesis includes ammonia-catalyzed reactions of tetraethylorthosilicate, TEOS ($Si(OR)_4$ with $R = C_2H_5$), with water in low-molecular-weight alcohols. This method enables the synthesis of spherical and monodisperse silica nanoparticles with the range in size from 5-2000nm and low polydispersity. The density of the silica particles is on the order of $\rho = 1.8 g/cm^3$. Compared to polystyrene spheres ($\rho = 1.05 g/cm^3$) sedimentation effects have a large influence on the arising structure of the suspensions. Thus the silica spheres must be small enough (below 100nm in diameter) so that the Brownian motion of the particles counteracts sedimentation.

The general and simplified reaction scheme of the silica synthesis includes a hydrolysis reaction

$$Si(OR)_4 + H_2O \xrightarrow{OH^-} (OR)_3Si(OH) + ROH \tag{4.1}$$

to produce the single-hydrolyzed TEOS monomer $(OR)_3Si(OH)$. Subsequently, this intermediate reaction product condenses to eventually form silica

$$(OR)_3Si(OH) + H_2O \rightarrow SiO_2 \downarrow + 3ROH. \tag{4.2}$$

This technique was refined to produce specifically particles of about 100nm with a low polydispersity in size. The precise diameter of the particles was determined by the dynamical light scattering technique introduced in sec. 3.2.2.

The synthesized particles carry weakly acidic silanol groups ($Si-OH$) on the surface which partly start to dissociate in a deionized water environment, leaving a population of spheres which have a slightly negative surface charge with narrow limits. Different to particles with acidic end groups the degree of dissociation of the silanol end groups is in most cases not high enough to cause a strong repulsive interaction. Since the driving force for the crystallization in charge stabilized colloids is due to electrostatic interaction, crystals are expected to be formed when the surface charge density is high enough. Surface groups of particles with acidic end groups are fully dissociated with their maximum possible surface charge in the deionized state. To influence the interaction one can add an electrolyte, e.g. NaCl, to screen the particles charge and reduce the interaction. In a silica system with silanol surface groups, a special mechanism exists to control the dissolubility. It is possible to control the degree of dissociation and the surface charge density with sodium hydroxide (NaOH). This mechanism is well known from literature [144–148] and is shown schematically in Fig. 4.1.

The principle mechanism by which silica surfaces acquire a charge in contact with water is the dissociation of silanol surface groups following the reaction (Fig. 4.1 (a)):

$$SiOH \rightarrow SiO^- + H^+. \tag{4.3}$$

In this case the resulting surface charge is not high enough to cause a strong electrostatic repulsion. Further protonation of the uncharged groups is expected at high pH values. These conditions can be achieved by adding sodium hydroxide to the suspension [144]. In this case silica particles charge up following the reaction:

$$SiOH + NaOH \rightarrow SiO^- + Na^+ + H_2O. \tag{4.4}$$

Thus the system stays deionized but the counter ion species changes from H^+ to Na^+ (Fig. 4.1 (b)) until its maximum surface charge is achieved (Fig. 4.1 (c)). Past the equivalence point all silanol groups are used up and the electrolyte concentration increases due to excess NaOH (Fig. 4.1 (d)). Both processes are reflected in conductometry by a decrease of conductivity followed by an increase. With the increasing surface charge density (bare charge) also the effective charge Z^* increases. The increase, however is governed by counter-ion condensation. Thus Z^* first increases linearly, then sublinearly and it finally saturates irrespective of the bare charge Z_{bare} [25]. After complete dissociation, the excess NaOH induces a strong screening and the interaction decreases again, while only minor changes are observed for Z^*. Thus a re-entrant crystallization behavior (liquid $\rightarrow$ crystalline $\rightarrow$ liquid phase transition) with increased NaOH concentration is expected.

The corresponding phase behavior in dependence of both control parameters, the particle number density n and NaOH concentration c_{NaOH}, is introduced in the next section 4.2 including a characterization in terms of elastic properties and growth behavior. Elastic properties by means of the shear modulus G give access to the effective charge Z^*. In addition the chemical potential difference $\Delta\mu$ between the metastable melt and stable solid state as an important measure for the degree of undercooling is determined via growth measurements. Its knowledge is not only indispensable for the interpretation of the behavior of the short-range order in the non-equilibrium state, but also for nucleation analysis within the framework of the classical nucleation theory (CNT).

4.2 Phase behavior and characterization

Charged colloidal systems have substantial experimental advantages in comparison to metals for studying the solid-liquid phase transition, since the particle interaction can be tuned in a wide range by varying experimental variables such as the particle number density, the electrolyte concentration and the effective surface charge density of the particles. Phase behavior of colloidal silica suspensions was investigated by Yamanaka et al. [144,145,147]. They analyzed the influence of NaOH concentration on the surface charge of silica spheres (120nm in diameter) stabilized by ammonium hydroxide. Nevertheless, these results lack systematical information on the phase formation and cover only the charging-up process at different NaOH concentration as described in sec. 4.1.

Within the present PhD thesis, two related colloidal suspensions were available for analyzing the model character of colloids. The analyzed systems consist of silica particles with a diameter of 84nm (Si84) and 77nm (Si77), respectively. Both systems are suspended in ultra pure water. The first system (Si84) had a maximum concentration of $n = 113 \mu m^{-3}$ in the stock suspension, corresponding to a volume fraction of $\Phi = 0.035$. It was used to analyze the short-range order within the first measurement campaign at HASYLAB (DESY). Due to improvements of the particle synthesis, a further silica system with a similar particle diameter of 77nm but with a higher particle concentration ($n = 224 \mu m^{-3}$ in the stock suspension, corresponding to a volume fraction of $\Phi = 0.053$) was available for the second measurement campaign. In combination with a detector (PILATUS100K) of advanced time-resolution, the investigations were focused on the nucleation behavior. The Si77

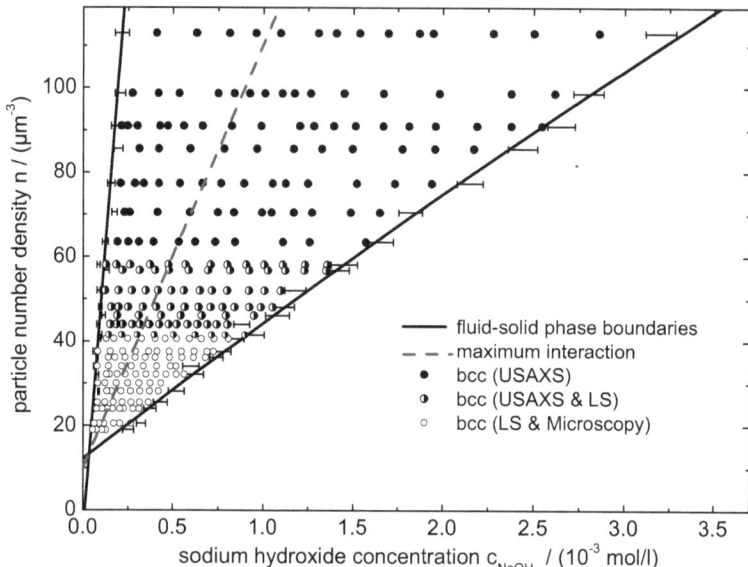

Figure 4.2: Phase diagram of Si84 in dependence on the particle number density and the sodium hydroxide concentration as determined by USAXS (black dots) and light scattering (white circles). Both measurement techniques show an overlapping regime with coinciding results for n between $40\mu m^{-3}$ and $60\mu m^{-3}$. The black lines are guides to the eyes for the solid-fluid phase boundaries with an uncertainty marked by the error bars. The dashed red line outlines the region of maximum interaction.

system shows qualitatively similar behavior concerning the phase diagram, elastic properties and growth behavior as the Si84 system. The quantitative differences and accessible ranges of particle number densities for the different experimental methods are summarized in table 4.1 at the end of this section. The evaluation and discussion of the characterization parameters are presented for the Si84 system exemplarily in detail in the following.

For the Si84 system the phase diagram was determined systematically in the range of particle number densities between $n = 18.0 \mu m^{-3}$ and $n = 113.0 \mu m^{-3}$ at various NaOH concentrations reflecting the re-entrant crystallization behavior with increased NaOH concentration shown in Fig. 4.2. At large n the structure of the silica suspension was studied by USAXS, at lower n by static light scattering. Intensity distributions in dependence of the scattering vector q were used to identify the structure and, in case of crystallizing samples, determine the n-dependent lattice constant. In the particle number range between $n = 42.0 \mu m^{-3}$ and $n = 58.0 \mu m^{-3}$ an overlap of the results is observable reflecting the complementary character of both scattering methods, light and X-ray scattering, respectively. The irradiation source for light scattering experiments had a wavelength of 532 nm and that of X-rays 0.138nm.

The suspension crystallizes in a bcc structure up to largest n. A crystalline phase is observed for number densities above $n = 18.0 \mu m^{-3}$. The crystalline region shows two fluid-solid phase boundaries. To the left the boundary is reached by charging the particles up at still deionized conditions. To the right it is reached by increased screening. The horizontal error bars denote the extension of the coexistence region. The solid lines are guides to the eyes dividing the crystalline bcc from the fluid phases. The dashed red line indicates the region of maximum interaction where all silanol groups on the particle surface are used up for the charging-up reaction. The conditions of maximum interaction are observed in measurements of the elasticity, which are discussed later in detail.

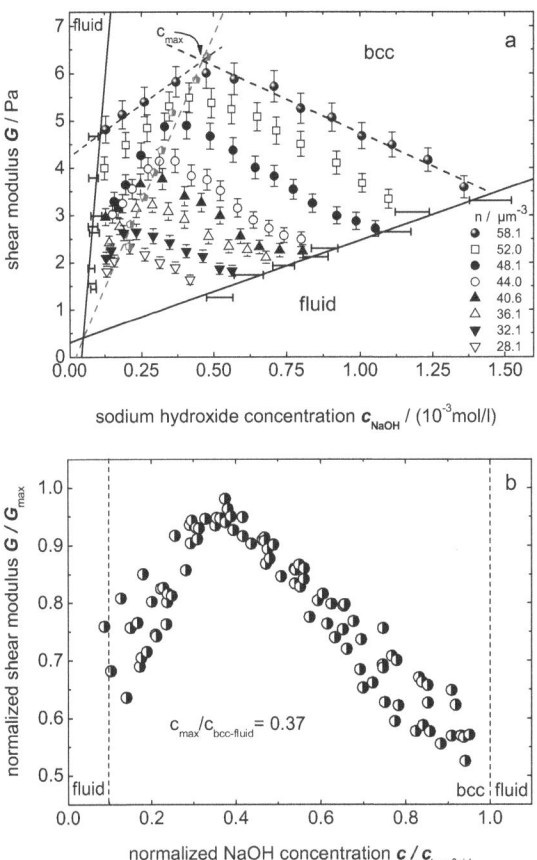

Figure 4.3: Shear modulus behavior of the colloidal silica crystals of Si84 systems in dependence on the concentration of added NaOH. a) For various n the shear modulus first increases up to a maximum value and then decreases. The dashed lines are fits to extrapolate the equivalence point [149] (demonstrated for $n = 58.1 \mu m^{-3}$ only). The maximum values at each particle number density can be described by a straight line representing the maximum interaction between the particles. b) Shear moduli normalized by the maximum shear modulus vs. the NaOH concentration which is again normalized by the concentration at the bcc-fluid phase boundary. All data superpose on a single master curve, where $\frac{c_{max}}{c_{bcc/fluid}} = 0.37$ holds irrespective of n, indicating that indeed both the charging and the following screening process scale with particle concentration.

The phase behavior of systems with spheres of fixed surface charge is already discussed in sec. 2.4 where Sirota *et al.* determined one of the first experimental phase diagram in dependence of the volume fraction and the electrolyte concentration derived systematically by Small Angle X-ray scattering [36]. In contrast to systems of fixed charge a re-entrant crystallization is observed as a function of either c_{NaOH} or n. At increasing particle number density under completely deionized conditions the charge is too small to cause crystalline order. Increasing the charge induces increased interactions and thus crystallization. At constant c_{NaOH} an increase of n reduces the amount of NaOH available per silanol group and thus first causes a decrease of screening but then a decrease

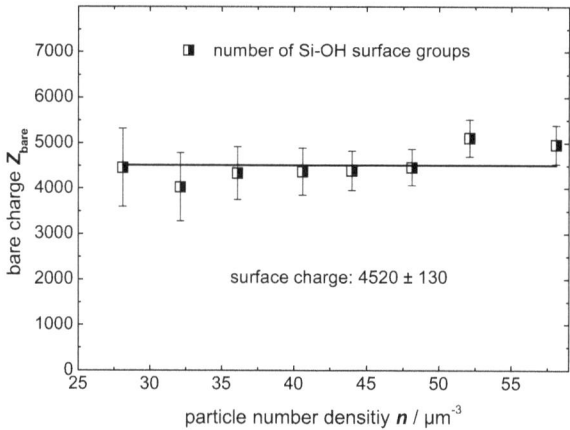

Figure 4.4: The conversion of the charge density values for Si84 at each concentration results in the mean number of silanol groups on each particle surface of $Z_{\text{bare}} = 4520 \pm 130$ (bare charge of a particle).

of the charge density. Thus re-entrant behavior here is given in two variables. This is a novel observation and different to the re-entrant behavior previously observed for polymer particles of constant bare charge. There, upon increasing the particle concentration at deionized conditions, the interaction was first increased then decreased, then increased again. The decrease was attributed to an intermediate increase of screening due to the added particle counter ions outweighing the decrease of inter-particle distance [150].

To pursue the phase behavior further, systematic measurements of the shear modulus were performed using the multi-purpose light scattering instrument (see section 3.2.3). With this method the shear moduli (G) were determined in a particle number density range between $n = 28.1 \mu m^{-3}$ up to $n = 58.1 \mu m^{-3}$ where homogeneous nucleation is the dominant nucleation process. At every particle number density the shear moduli were measured in dependence on the added amount of base c_{NaOH} under an inert Argon atmosphere. Figure 4.3 (a) shows the shear modulus behavior in dependence on the NaOH concentration for various particle number densities. With increasing NaOH concentration the shear modulus first increases approximately linearly due to the charging up reaction of the silica particles. The maximum shear modulus indicates the maximum interaction between the particles. Further addition of excess NaOH linearly decreases the shear modulus due to the screening of the particles.

In order to obtain the equivalence point, the so called shear modulus titration [149] can be applied. Fig. 4.3 (a) demonstrates the evaluation process for $n = 58.1 \mu m^{-3}$ only. The dashed lines are fits to extrapolate the equivalence point [149] at the corresponding NaOH concentration that is in turn associated to the state of maximum interaction. The maximum values of the shear moduli at different concentration can be described by a straight line (see Fig. 4.3 (a)). Furthermore, all curves have a similar shape. This in fact is expected, if first the charging procedure scales with both n and c_{NaOH} and second a polycrystalline morphology is retained [151]. To check the latter condition the shear moduli were normalized by their maximum value and the NaOH concentration by its value at the upper phase boundary. All measurement series covering a particle number range between $n = 28.1 \mu m^{-3}$ and $n = 58.1 \mu m^{-3}$ collapse on a single master curve, indicating the self similarity of the charging process during this shear modulus titration experiment [149]. Moreover, $\frac{c_{\text{max}}}{c_{\text{bcc/fluid}}} = 0.37$ does not change by varying n, indicating that indeed both the charging and the following screening process scale with particle concentration (see Fig. 4.3 (b)). Here, c_{max} and $c_{\text{bcc/fluid}}$ denote the NaOH concentration at maximum interaction and at the right solid-fluid phase boundary, respectively.

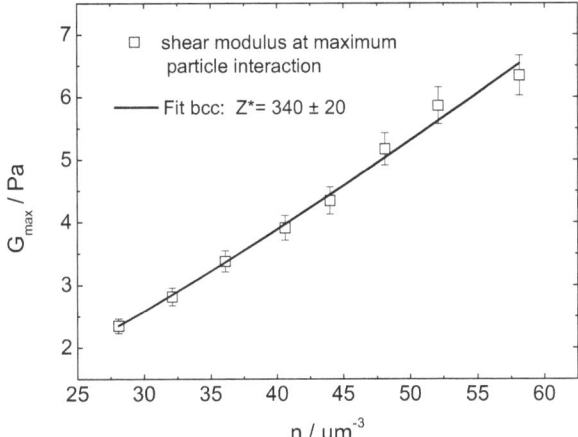

Figure 4.5: Maximum values of the shear modulus in dependence on the particle number density. The solid line is the best one parameter fit with the effective charge as the only free fit parameter. The effective charge is constant within the experimental uncertainty and the linear least square fit yields a value of $Z^* = 340 \pm 20$ for the effective charge.

The master behavior of the shear modulus is an indication of that $\frac{c_{max}}{c_{bcc/fluid}} = 0.37$ does not change as a function of n. This fact allows the region of maximum interaction to be approximated linearly for higher n, where the elastic properties of the silica system are not accessible experimentally. The region of maximum interaction is outlined by the dashed red line in Fig. 4.3 (a) and was transferred to the phase diagram in Fig. 4.2. The maximum interaction can be identified with the equivalence points determined by using the technique of graphical direction lines in the behavior of increasing and decreasing shear modulus crossing the maximum value.

At maximum interaction the added NaOH base is completely used up for charging up the silica particles to the maximum possible charge. Therefore, the maximum surface charge density σ_a can be calculated from the added amount of base as [144]:

$$\sigma_a = \frac{10^{-3}}{3} N_A e a \frac{c_{NaOH}}{\Phi} \quad (4.5)$$

where N_A is Avogadro's number, e the elementary charge (in C), a the particle radius (in cm) and $\Phi = (4/3)\pi a^3 n$ the volume fraction of the silica particles. The conversion of the charge density values at each concentration results in the mean number of silanol groups on each particle surface of $Z_{bare} = 4520 \pm 130$ (bare charge of a particle see Fig. 4.4). Note that the exact unit of particle charge should be given in units of elementary charge in the narrow sense. Here, the charge number Z per particle is used for describing the charge.

The corresponding effective charge for a polycrystalline morphology can be determined from the shear moduli. The maximum shear moduli in dependence of the particle number density n are shown in Fig. 4.5. The best one parameter fit using Eq. (2.67) yields $Z^* = 340 \pm 20$. Similar to the case of polymer particles under deionized conditions Z^* is constant as a function of n within experimental uncertainty [152].

In addition to elastic measurements also growth measurements were conveniently carried out at low particle number densities in a range between $n = 18\mu m^{-3}$ up to $n = 37\mu m^{-3}$. Here polarization microscopy was applied to observe the subsequent growth of heterogeneously nucleated bcc wall crystals within a flat rectangular cell ($d = 1mm$ wall-to-wall distance). In this geometry wall crystal

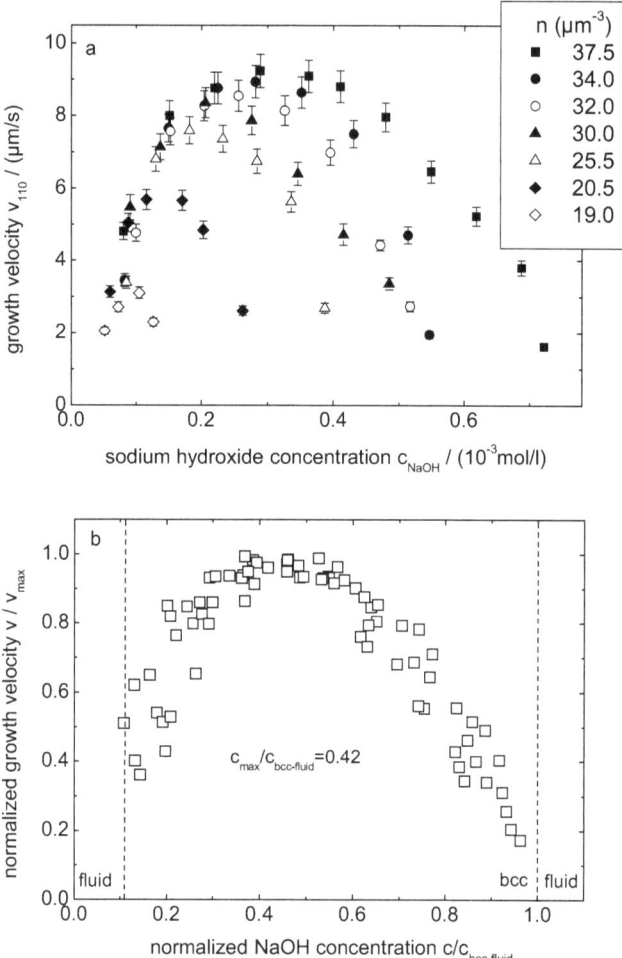

Figure 4.6: Measured growth velocities of Si84 versus amount of added base for different particle concentrations n. a) At small c_{NaOH} the velocities increase, then go through a flat maximum and decrease again for large base concentrations. b) Growth velocities normalized by the maximum values vs. the NaOH concentration which is again normalized by the concentration at the bcc-fluid phase boundary. All data superpose on a single master curve, where $\frac{c_{max}}{c_{bcc/fluid}} = 0.42$ holds irrespective of n, indicating that indeed both the charging and the following screening process scale with particle concentration.

growth dominates the solidification up to $n = 37 \mu m^{-3}$. Wall crystal pictures were collected with a CCD-camera and further evaluated with image processing software. The computer controlled preparation circuit facilitates a synchronization of the beginning of the picture collection by the CCD camera and the abrupt stop of the pumping of the suspension through the flat sample cell by means of magnetic valves. The time-dependent microscopy experiments are described precisely in sec. 3.3. This method is restricted by the onset of homogeneous nucleation that was observed at $n \approx 37 \mu m^{-3}$.

The measurements of the growth velocity were performed at various fixed particle number densities

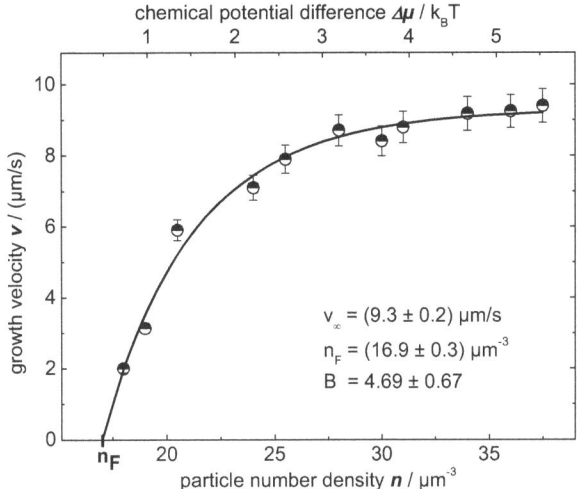

Figure 4.7: The growth velocity v_{110} for the Si84 system as function of the particle number density n at maximum interaction. The solid curve is a fit of a Wilson-Frenkel law (2.56) yielding $v_\infty = (9.3 \pm 0.2)\mu m s^{-1}$ and $B = (3.91 \pm 0.67)k_B T$. The chemical potential difference $\Delta\mu$ between solid and melt used in the upper x axes was derived using the conversion factor B. Nearby the phase boundary n_F the behavior of growth velocity shows a strong dependence on the particle number density. With increasing particle number density the growth velocity saturates [76, 88, 91].

as a function of the added amount of NaOH. In all cases a linear increase of the crystal dimension with time was observed (see Fig. 3.9) indicating reaction controlled growth. The results of the growth velocity as function of n and c_{NaOH} are given in Fig. 4.6 (a). Here the behavior of the growth velocity can be described by a convex parabolic curve. But again a good scaling is observed in a plot of the velocity normalized by the maximum velocity and the base concentration by that at the upper phase boundary as shown in Fig. 4.6 (b). The ratio $c_{max}/c_{bcc/fluid} = 0.42$ appears only slightly above the corresponding value extracted from the shear modulus measurements.

Fig. 4.7 shows the growth velocity at maximum interaction in dependence on the particle number density n. Nearby the phase boundary n_F between the fluid and the solid state the behavior of growth velocity shows a strong dependence on the particle number density. With increasing particle number density the growth velocity saturates [76, 88, 91]. The flattening is understandable considering that reaction controlled growth possesses a limiting velocity. That sort of behavior can be well described with the Wilson Frenkel law following the formalism of Aastuen *et al.* as introduced in section 2.6.5 in detail. Here a fit of the Wilson-Frenkel law (2.56) of the experimentally determined growth velocities at maximum interaction with the limiting velocity v_∞ at infinite undercooling and B as the only free parameters are obtained. The result is shown in Fig. 4.7. The best fit to the experimental data gives values of $v_\infty = (9.3 \pm 0.2)\mu m/s$ for the limiting growth velocity and $B = (4.69 \pm 0.67)k_B T$. The good fit verifies the formulation of the potential difference driving the phase transition. Therefore it is possible to interpret the fit parameter B as a conversion factor between the reduced particle number density $((n - n_F)/n_F)$ and the chemical potential difference $\Delta\mu$ between the solid and the melt. This is scaled in the upper x axes of Fig. 4.7.

The difference of the chemical potential between the metastable fluid and the stable solid state as a function of the particle number density n shows a nearly linear dependence on n, which can be extrapolated at higher particle number densities, where the growth velocity is not accessible experimentally. These results are shown in Fig. 4.8. This evaluation method makes it possible to obtain

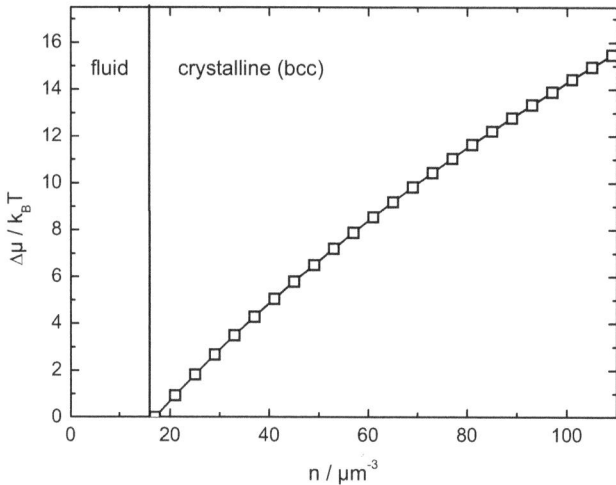

Figure 4.8: Chemical potential difference between the metastable fluid and the stable solid state as a function of the particle number density n for the Si84 sample determined from growth measurements followed by an evaluation with the Wilson-Frenkel law. Over the investigated concentration range ($n = 18\mu m^{-3}$ and $n = 37\mu m^{-3}$), $\Delta\mu$ shows a nearly linear dependence on n, which can be extrapolated at higher particle number densities.

particle system	**Si77**	**Si84**
diameter $2a$ / nm	77	84
polydispersity p / %	8	8
maximum volume fraction Φ_{max}	0.0535	0.0355
effective charge Z^*	260± 10	340± 20
bare charge Z_{bare}	2240± 260	4520±130
range of investigated particle number densities n / μm^{-3}	30.0 - 224.7	18.0 - 113.0
lattice constants g / nm	210 - 360	260 - 420
melting point n_F / μm^{-3}	30	18
range of particle number densities in which the shear modulus was measured n / μm^{-3}	38.5 - 68.0	28.1 - 58.1
range of particle number densities in which the growth velocity was measured n / μm^{-3}	30.0 - 52.0	18.0 - 37.0

Table 4.1: Overview of important properties of the analyzed silica suspensions Si77 (particles with 77nm in diameter) and Si84 (particles with 84nm in diameter) including the particle diameter $2a$, the maximum volume fraction Φ_{max}, the polydispersity p, the bare charge number Z_{bare} of the soluble surface charge density, the effective charge number Z^* at maximum interaction, the melting point n_F and the lattice constants g. In addition to the investigated particle number density n also the ranges of the particle number densities are given in which the shear modulus and the growth velocity were measured. Note the higher melting point of Si77 at $n_F = 30.0 \mu m^{-3}$ due to the lower surface and effective charge at comparable particle size in the Si84 system.

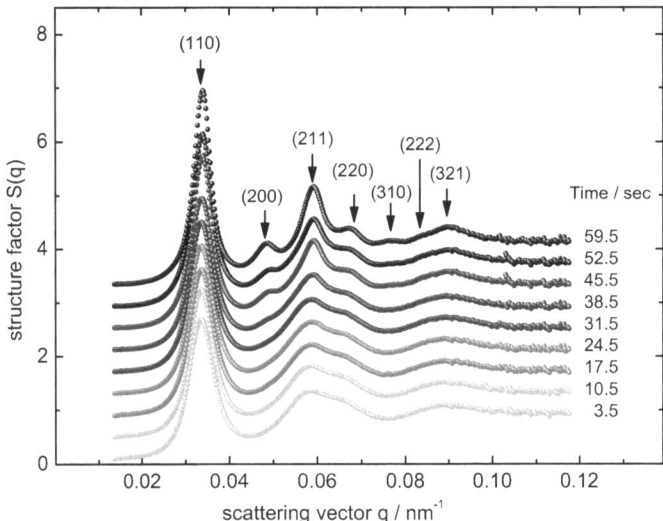

Figure 4.9: Time dependent static structure factors $S(q)$ for a colloidal suspension at $n = 113 \mu m^{-3}$ and maximum interaction that was shear melted to suppress the crystallization. By doing this, the colloidal melt is set into a metastable state (lower curves) and subsequently crystallizes into a bcc structure (upper curves). $S(q)$ are shifted for clarity and time increases from bottom to top. Miller indices are indicated, clearly identifying the bcc structure of the polycrystalline solid formed.

the chemical potential difference $\Delta \mu$ as a function of n at maximum interaction in the deionized state. Basically, this method can also be applied at interactions different from the maximum one. In that case, the undercooling becomes a function of n as well as the excess NaOH concentration c_{NaOH}:

$$v = v_\infty \left[1 - \exp\left(\frac{\Delta \mu(n, c_{\text{NaOH}})}{k_\text{B} T} \right) \right] \tag{4.6}$$

For the chemical potential difference again a master curve can be calculated from the results of growth behavior in the same way as in the case of shear moduli and growth velocities.

The entire set of characteristic properties of colloidal systems, which are introduced in the present section and summarized in table 4.1 is a necessary background for further investigations of the crystallization process. The chemical potential difference between the metastable fluid and the stable solid state is an important parameter to describe the degree of metastability. It plays an important role in the following section where the short-range order in the metastable state is discussed. With the undercooling parameters a major prerequisite is also gained for the extraction of interfacial tensions from the nucleation experiments within the framework of CNT.

4.3 Short-range order in charged sphere colloids

The knowledge of short-range order in a liquid is of fundamental importance for understanding the nucleation behavior of solid phases from the melt. The first attempt to compare short-range order of alkali metals like Rb, K, Cs in the liquid state and colloidal fluids under equilibrium conditions was reported by Cotter and Clark [153]. The present work addresses the short-range order far away from equilibrium in the liquid phase of colloidal systems in comparison to metals.

For metals solidification under non-equilibrium conditions can be achieved by e.g. rapid quenching or undercooling experiments. In the latter experiments, a melt is undercooled at low or moderate cooling rates below its melting temperature. The driving force for crystallization scales with the difference of Gibbs free energy of solid and liquid state and increases with increasing undercooling. Colloidal systems undergo isothermal phase transitions without the temperature as the crucial parameter for the phase transition. Suppression of crystallization is achieved by shear forces which are able to destroy the crystal to leave the system in a metastable liquid state. After cessation of shear, the suspension is in a metastable state of fluid order, equivalent to the metastable state of an undercooled metallic melt, and starts to crystallize after a characteristic induction time for nucleation. Fig. 4.9 shows the time evolution of the structure factor of the colloidal system from the non-equilibrium liquid state (lower curve) to the stable solid phase (upper curve) at a particle number density $n = 113 \mu m^{-3}$. The time interval between each curve corresponds to 7s due to the integration interval of 3.5s and the detector read out time of 3.5s. Fig. 4.9 demonstrates that for times $< 25s$ there is no change of the structure factor. This assures that especially the first structure factor taken after 3.5s corresponds to a metastable liquid state. For longer times a structural transition into a stable bcc phase is observed. This behavior is representative for the analyzed system in the metastable state over the accessible range of the particle concentration. Deviations from equilibrium are described by the difference of the chemical potentials between the metastable melt and stable solid in colloidal suspensions and by the undercooling below the melting temperature of metals, respectively.

Investigations of the short-range order as a function of undercooling on charged colloids are rare in literature. The colloidal silica system investigated in the present work belongs to the group of strongly charged colloids. Systematic studies of the short-range order by measurements of the structure factor $S(q)$ both in the metastable and stable liquid state over the entire accessible phase diagram are subject of the present thesis and will be presented in the following section.

4.3.1 Short-range order of charged colloidal melts at different interaction strengths

The static structure factor of the metastable colloidal melt was investigated at various particle number densities in dependence of the NaOH concentration. The analyzed colloidal silica system provides two different mechanisms to influence the interaction. On the one hand, the strength of interaction can be increased with increasing particle number density. A re-entrant phase behavior can be achieved by varying the sodium hydroxide concentration on the other hand. In the latter case, the strength of interaction reaches its maximum value at a characteristic NaOH concentration for each particle number density.

Fig. 4.10 shows the structure factors for different particle number densities between $n = 46.1 \mu m^{-3}$ and $n = 113 \mu m^{-3}$ of the metastable liquid state at maximum interaction (red line in Fig. 4.2). With increasing particle number density the Bragg peaks shift to higher q-values as expected due to decreasing next-nearest neighbor distances. More surprising is the presence of the asymmetry in the second oscillation of $S(q)$. Over the entire accessible particle number densities an asymmetric second oscillation arises in the structure factor. The asymmetry in the second oscillation of $S(q)$ becomes more pronounced with increasing interaction which is associated with an increase in n.

The structure of the stable solid phase is clearly identified to be of a bcc type (Fig. 4.9). Obviously, it differs from the structure in the metastable state. An attempt to clarify the local order in metastable colloidal suspensions is provided in the next section.

4.3.2 Determination of the short-range order in the meta- stable state

Scattering reveals information in reciprocal space. To obtain real-space information, theoretical models and simulations of $S(q)$ are required, since scattering measurements deal with positive values of intensities that lack phase information. Here the formalism of Simonet [154, 155] is used to determine the short-range order in the regime of large q-vectors. This method assumes one dominant type of isolated less tightly bound structural units in the melt. In addition long-range inter

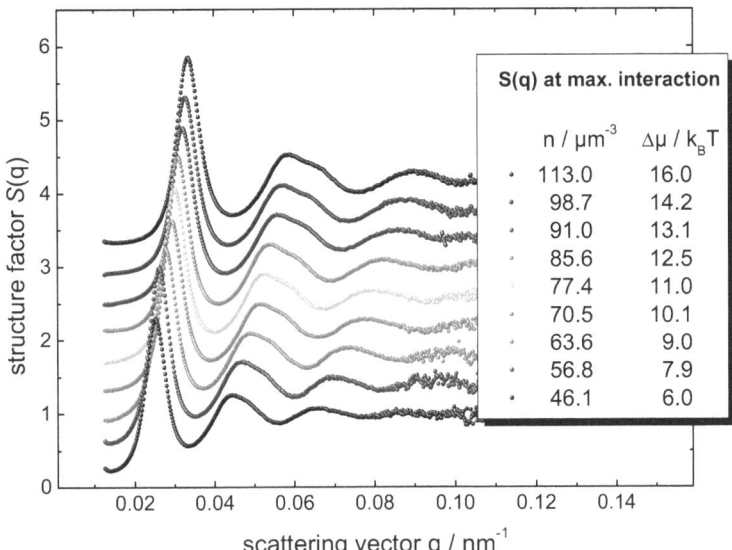

Figure 4.10: Static structure factors $S(q)$ of a colloidal silica suspension in the metastable state at maximum interaction (red line in the phase diagram Fig. 4.2) in dependence of n. Increasing interaction is associated with an asymmetry in the second oscillation of $S(q)$ which becomes more pronounced with increasing particle number density n.

cluster contributions to the scattered intensity are neglected and only intra cluster contributions are considered. This simplification is justified in the regime of large q values where the contributions of tightly bound longer inter cluster distances are damped out by thermal motions described by the Debye-Waller factor. The structure factor at large q is given by

$$S(q) = 1 + \frac{c}{Nb^2} \sum_{i,j(i \neq j)}^{N} b_i b_j \frac{\sin(q\langle r_{ij}\rangle)}{q\langle r_{ij}\rangle} \exp\left(-\frac{2q^2 \langle \delta r_{ij}^2 \rangle}{3}\right). \quad (4.7)$$

Here, N corresponds to the number of particles in a structural unit, $\langle r_{ij} \rangle$ the mean distance between the particles i and j and $\langle \delta r_{ij}^2 \rangle$ the mean thermal variation that determines the Debye-Waller factor $\exp\left(-2q^2\langle \delta r_{ij}^2\rangle/3\right)$. b_i denotes the scattering amplitude of particle i and b^2 is the average of the squares of the scattering amplitudes of all particles in one structural unit. This simulation method depends on three free parameters. These are the shortest mean distance $\langle r_0 \rangle$ of the particles, its mean thermal variation $\langle \delta r_0^2 \rangle$ and the fraction of particles c organized in each structural unit. For a given structure, all particle distances $\langle r_{ij} \rangle$ can be calculated from $\langle r_0 \rangle$ and the mean thermal variations $\langle \delta r_{ij}^2 \rangle$ from $\langle \delta r_0^2 \rangle$ at the shortest particle distance assuming $\langle \delta r_{ij}^2 \rangle = \langle \delta r_0^2 \rangle \langle r_{ij} \rangle^2 / \langle r_0 \rangle^2$. The three free parameters are adjusted such that a good agreement with the experimental $S(q)$ is obtained especially at large q values assuming following structures of the structural units of the liquid: body centered cubic (bcc), face centered cubic (fcc) or equivalently hexagonal close-packed (hcp) clusters as well as icosahedral and dodecahedral short-range order of the clusters were considered.

Simulated structure factors $S(q)$ for a colloidal silica suspension in the metastable liquid state at maximum interaction and at particle concentration $n = 46.1 \mu m^{-3}$ are shown in Fig. 4.11. For this system the deviation from the equilibrium state is characterized by $\Delta \mu = 6 k_B T$. The silica suspension forms a bcc phase in the solid state. If a short-range order of bcc structure is assumed, neither the position nor the shape of the asymmetric oscillation of $S(q)$ is described. The fit for a short-range order consisting of fcc clusters describes the experimental data better in the range of

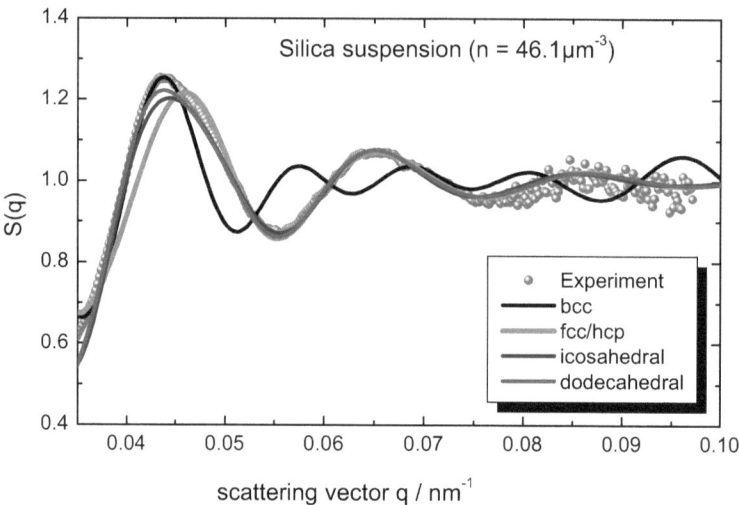

Figure 4.11: In comparison to the experimentally determined structure factor also simulated structure factors $S(q)$ of a colloidal silica suspension in the metastable state near the maximum interaction and a particle concentration $n = 46.1 \mu m^{-3}$ are shown. Here the formalism of Simonet et al. [154, 155] is used to analyze the short-range order in the regime of large q-vectors. The silica suspension forms a bcc phase in the solid state. If a short-range order of bcc structure is assumed, neither the position nor the shape of the asymmetric oscillation of $S(q)$ is described. The fit for a short-range order consisting of fcc clusters describes the experimental $S(q)$ better at higher oscillations, but not the shape at lower q-values. In contrast, for an icosahedral short-range order a good fit of the experimental data is achieved which becomes even better, if larger dodecahedral aggregates of the similar five-fold symmetry are assumed. The fit for a short-range order consisting of hcp clusters resembles that of fcc clusters and is therefore not shown explicitly.

r_0/nm	$\delta r_0^2 / nm^2$	c / %	type of short-range order
290	475	99	bcc
398	450	99	fcc
315	487	99	icosahedral
319	435	99	dodecahedral

Table 4.2: Fit parameters of the simulation procedure of Simonet for the silica system Si84 at constant particle number density $n = 46.1 \mu m^{-3}$ assuming an fcc, bcc, icosahedral and dodecahedral short-range order.

higher q values, but not the shape of the asymmetric oscillation. In contrast, assuming an icosahedral short-range order leads to a much better description of the measured $S(q)$. It becomes even better, if larger dodecahedral aggregates that are also characterized by a five-fold symmetry are assumed. This may indicate that a short-range order consisting of larger polytetrahedral aggregates such as dodecahedra prevails in the melt.

The corresponding fit parameters according to the model of Simonet are summarized in table 4.2. The fit for a short-range order consisting of hcp type units resembles that for fcc aggregates and is therefore not shown explicitly. The model of Simonet et al. depends on the distance between the particles in an aggregate that is independent on the layer arrangements. While the particle distances stay the same for both structures, the layer arrangements differ between the fcc (...ABCABC...) and hcp (...ABAB...) close-packing.

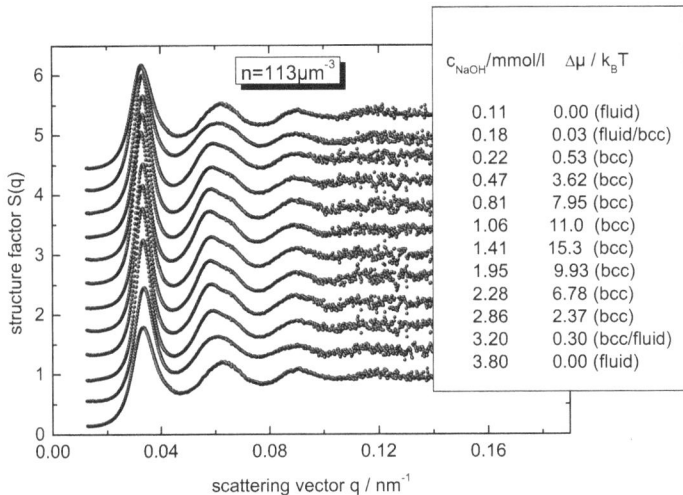

Figure 4.12: Experimentally determined static structure factors $S(q)$ of a colloidal silica suspension at constant particle number density $n = 113 \mu m^{-3}$ in dependence of the sodium hydroxide concentration c_{NaOH}. The interaction strength and consequently the asymmetry in the second oscillation of the structure factor first increases up to about $c_{NaOH} = 1.4 mmol/l$ (corresponding to the state of maximum interaction) and then decreases again with increasing c_{NaOH}. Crossing the phase boundary into the stable fluid state, the asymmetry disappears completely.

The origin of the asymmetric second oscillation is explained when considering the individual contributions of each inter atomic distance within an aggregate of fcc or hcp in comparison to an icosahedral type of structure. In fcc or hcp aggregates, the distance of the central particle to the particles on the shell r_0 and the nearest neighbor distance on the shell are the same but differ by about 5% in icosahedral aggregates. The slight difference of the particle distances is responsible for the asymmetry in the second oscillation of $S(q)$. The shoulder can therefore be considered as an indication for icosahedral short-range order. The model of Simonet certainly neglects the wealth of imperfect aggregates of five fold symmetry [70] which are also likely to be present. But it is sufficient to deliver information on the dominant type of short-range order in the system.

For pure metallic melts with compact local order and isotropic bonding Frank [2] hypothesized an icosahedral short-range order independent of the structure of the corresponding solid phases. Frank's hypothesis was used to explain the large undercoolings of pure metallic melts observed by Turnbull [1, 156] such that the fivefold symmetry of icosahedral short-range order must be broken before a crystal with its translational symmetry can be formed. Five-fold symmetry belongs to the crystallographically 'forbidden' point group due to its incompatibility with translational invariance. The presented results on the short-range order in undercooled charge stabilized colloids reveal a structure of icosahedral symmetry quite similar as measured in undercooled melts of pure metals despite the fact that the corresponding solid phase in the colloidal system is of bcc type.

4.3.3 Short-range order of charged colloidal melts near the solid-fluid phase boundary

For investigations of the short-range order at a solid-fluid phase boundary at maximum interaction in principle one has the possibility to go to the lowest particle number densities below $n = 18.0 \mu m^{-3}$, where the system stays fluid (see Fig. 4.2). In this particle concentration regime, the structure is

c_{NaOH} /$10^{-3} mol/l$	r_0/nm	$\delta r_0^2 / nm^2$	c / %	short-range order
1.06	232	143	99	dodecahedral
2.86	232	165	99	dodecahedral
3.30	230	233	99	icosahedral
3.80	224	173	80	fcc

Table 4.3: Fit parameters of the simulation procedure of Simonet for the silica system Si84 at constant particle number density $n = 113 \mu m^{-3}$ and variable NaOH concentrations assuming an fcc, bcc, icosahedral and dodecahedral short-range order.

accessible for measurements by light scattering only. This method however offers no possibility to measure the structure factor at higher oscillations, which include the necessary information concerning the short-range order. Higher oscillations are accessible to measure using the USAXS method, but this method is also limited with respect to the particle number density. Low particle number densities imply larger distances between the particles resulting to small scattering angles at which scattering signals are not detectable due to the beam stop. The lowest particle number density accessible for USAXS measurements is at about $n = 40 \mu m^{-3}$. At higher particle concentrations at maximum interaction no solid-fluid transition is expected. Theoretical predictions rather assume a transition to more dense packed crystal structures from bcc to fcc phase [26]. Glass-like phases with short-range order could also be possible at higher particle number densities rather than a fluid state [36, 37]. Furthermore, an experimental evidence for a bcc-fcc phase transition at high volume fractions was also provided by Sirota et al. [36].

Another possibility to analyze the short-range order at a solid-fluid phase transition is to follow the structure factor at constant particle number density, but variable sodium hydroxide concentrations. Fig. 4.12 shows static structure factors $S(q)$ of a colloidal silica suspension at constant particle number density $n = 113 \mu m^{-3}$ in dependence of the sodium hydroxide concentration c_{NaOH}, where a reentrant phase transition is observed. The degree of metastability is given by the chemical potential difference $\Delta \mu$ obtained by a two-parameter fit in terms of a Wilson-Frenkel equation (2.55) in dependence of the particle number density n and the NaOH concentration c_{NaOH} (4.6).

Crossing the left liquid-solid phase boundary (see the phase diagram in Fig. 4.2 or the $S(q)$ from top to bottom in Fig. 4.12) from the fluid into the solid state at $c_{NaOH}= 0.18 mmol/l$, the structure factors of the metastable melt are changing their shape especially concerning the second oscillation from a symmetric to an asymmetric shape. With increasing NaOH concentration up to $c_{NaOH}= 1.41 mmol/l$ at maximum interaction, the 'shoulder' of $S(q)$ becomes more pronounced. At further addition of NaOH with decreasing strength of interaction due to the screening effect, the asymmetry becomes less pronounced until it disappears completely at $c_{NaOH}= 3.80 mmol/l$.

To reveal the corresponding structure, the same formalism as introduced in sec. 4.3.2 and already presented in the previous section was used. These results are shown in Fig. 4.13. For a better overview, only the structure factors at the following NaOH concentrations are shown: $c_{NaOH}= 1.06 \cdot 10^{-3} mol/l$ (near the maximum interaction), $2.86 \cdot 10^{-3} mol/l$ (near the solid-fluid phase boundary), $3.20 \cdot 10^{-3} mol/l$ (bcc-fluid phase boundary) and $3.80 \cdot 10^{-3} mol/l$ (stable fluid). The best description of the simulated structure factors is obtained for a dodecahedral short-range order in the metastable state near the maximum interaction ($c_{NaOH}= 1.06 mmol/l$) and at higher NaOH concentrations near the phase boundary ($c_{NaOH}= 2.86 mmol/l$). However, exactly at the phase boundary ($c_{NaOH}= 3.20 mmol/l$) an icosahedral short-range order consisting on simple 13 atomic icosahedral aggregates describes the experimental structure factor in the best way. Interestingly, as indicated already by the absence of the asymmetry in the second oscillation of $S(q)$, no icosahedral short-range order exists in the fluid state at high NaOH concentrations ($c_{NaOH}= 3.80 mmol/l$). A short-range order with fcc structure dominates the fluid in this regime. The corresponding fit parameters from the model of Simonet are summarized in table 4.3.

The behavior of the short-range order at the solid-fluid phase boundary is explained as follows. Through the addition of screening electrolyte, the interaction character is changed continuously

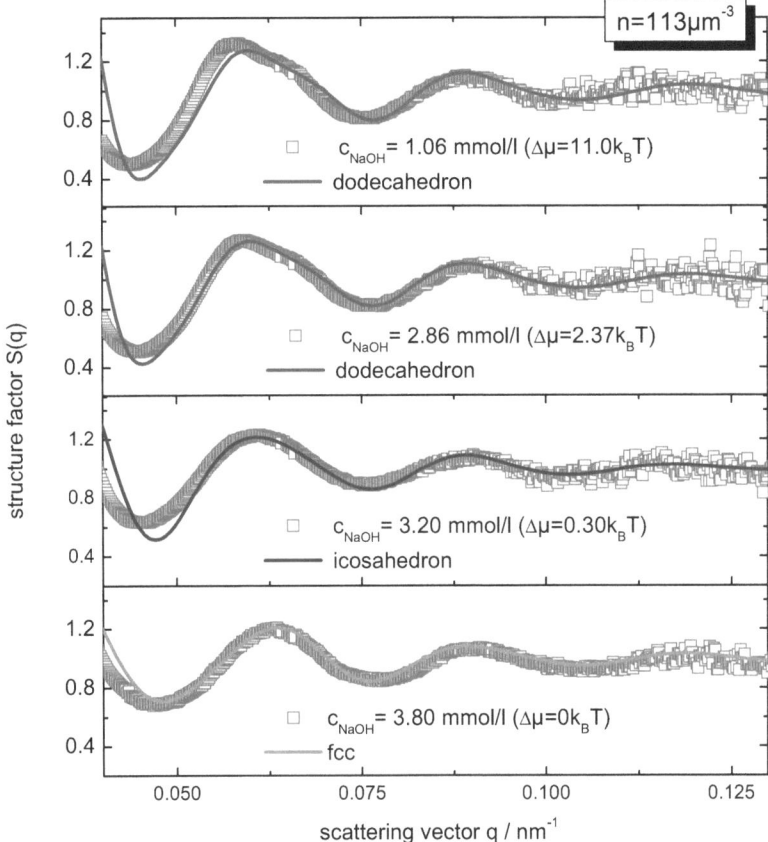

Figure 4.13: Simulations (lines) of the experimentally determined (squares) static structure factors $S(q)$ of a colloidal silica suspension at constant particle number density $n = 113 \mu m^{-3}$ in dependence of the sodium hydroxide concentration: $c_{\text{NaOH}}= 1.06, 2.86, 3.30$ and $3.80 \cdot 10^{-3} mol/l$. The simulated structure factors exhibit a dodecahedral short-range order in the metastable liquid state at maximum interaction ($c_{\text{NaOH}}= 1.06 mmol/l$) and at higher NaOH concentrations near the phase boundary ($c_{\text{NaOH}}= 2.86 mmol/l$). However, exactly at the solid-liquid phase boundary ($c_{\text{NaOH}}= 3.20 mmol/l$) an icosahedral short-range order consisting of smaller 13 atomic icosahedral aggregates, describes the experimental structure factor in the best way. A short-range order with fcc structure dominates the fluid at higher NaOH concentrations ($c_{\text{NaOH}}= 3.80 mmol/l$).

from soft to nearly hard sphere like behavior. In simulations using hard sphere potentials [67], no icosahedral short-range is observable. This leads to indications that the soft repulsive interaction is supposed to be responsible for the formation of icosahedral short-range order. In pure metals the asymmetry of the second oscillation of $S(q)$ is not only present in the undercooled state but also, even if less pronounced, above the melting temperature T_L. Holland-Moritz et al. [51] showed that independently of the degree of metastability, a Lennard-Jones system as Zr shows a short-range order of five-fold symmetry even at $T > T_\text{L}$. The icosahedral short-range order becomes more pronounced, if the temperature T is decreased. Similar conclusions as for the Zr melts were drawn from simulations of $S(q)$ of Fe, Ni and Co [51,59,157,158].

At this point it should be clearly pointed out that there is a fundamental difference in the phase transition behavior between charge stabilized colloids and metals. In metals the interaction potentials stay the same for the analyzed system even if the temperature is varied. When varying the temperature at constant pressure the entropic term of the Gibbs free energy:

$$G(T) = H(T) - TS(T), \tag{4.8}$$

at a phase is changed. With increasing temperature structures with a low degree of order will be favored. Therefore a state like the fluid state is stable at high T. For the liquid phase an increase of T will lead to a less pronounced short-range order, but due to the unchanged interaction potentials the general type of short-range order will not change as experimentally observed.

For the investigated colloidal system however, the interaction potential is changing by variation of the concentration of screening electrolyte and the temperature always stays constant even at solid-liquid phase transitions. The corresponding change of the enthalpic terms on the one hand may influence the type of short-range order of the fluid phase, on the other hand it also determines the phase stability as function of the screening electrolyte concentration.

4.3.4 Short-range order of charged colloidal melts in comparison to metals

The present section deals with a direct comparison of the short-range order in colloids and metals to discuss the model character of colloidal systems. Measurements of the structure factor both of fluid monodisperse colloids and molten pure metals are presented in this section.

Fig. 4.14 compares experimentally determined structure factors of the non equilibrium liquid state for a colloidal system measured by USAXS (a) with experimentally determined structure factors of Ni melt previously measured by neutron scattering at various undercoolings (b) [59]. In order to compare directly the structure factors of the colloidal suspension and the metal, the structure factors of the melts are plotted as function of the dimensionless wave vectors q/q_{max} with q_{max} the wave vector at the first maximum of $S(q)$. The deviations from equilibrium for the colloidal suspension are given by the chemical potential difference $\Delta\mu$ between the metastable fluid and stable solid. Corresponding driving forces in metallic systems can be determined via the Gibbs free energy difference $\Delta G = G_{\text{solid}} - G_{\text{liquid}} \propto \Delta\mu$, which is estimated by the linear approximation $\Delta G = \Delta S(T_L - T)$ with ΔS the entropy and T_L the melting temperature of the metal [51].

The diffraction measurements are strikingly similar and show up to four oscillations with decreasing intensity if q/q_{max} is increasing. In both systems the experiments reveal an asymmetric second oscillation of $S(q/q_{max})$ with a shoulder that becomes more pronounced with increasing deviations from thermodynamic equilibrium. This behavior is reflected in the range of particle number densities between $n = 46.1 \mu m^{-3}$ and $n = 113.0 \mu m^{-3}$ for the silica suspension and in the temperature range between $T = 1435K$ and $T = 1905K$ for the Ni melt. The shoulder is considered as an indication for icosahedral short-range order to be present in both physically different systems. Interestingly, in metals an icosahedral short-range order is also observed in the state above the melting temperature $T_L = 1726K$, although the shoulder in the second oscillation of $S(q/q_{max})$ is less pronounced due to high temperature which dominate the entropic term in the Gibb's free energy.

Independent of the different length scale in atomic and colloidal systems, the formalism proposed by Simonet [154] introduced in section 4.3.2 was used to analyze the short-range order in both systems. The simulated structure factors of the colloidal suspension and the metallic melt are compared with the experimental results in Fig. 4.15. Supposing a bcc-like short-range order in the undercooled melt fails in describing the data. A fcc type short-range order performs better but the shape and peak positions in the second oscillation do not agree with the experimental results where the significant asymmetry of a shoulder is observed. A good agreement with the measured data is achieved if icosahedral units of fivefold symmetry are assumed. It even becomes better assuming larger dodecahedral aggregates of equal five-fold symmetry. The best fit of the simulations to the experiments returns for icosahedral and dodecahedral ordering with fit parameters as compiled in table 4.4. The better agreement of the simulation for icosahedral clusters in comparison to fcc aggregates can be explained by considering the individual contributions of each inter-atomic distance

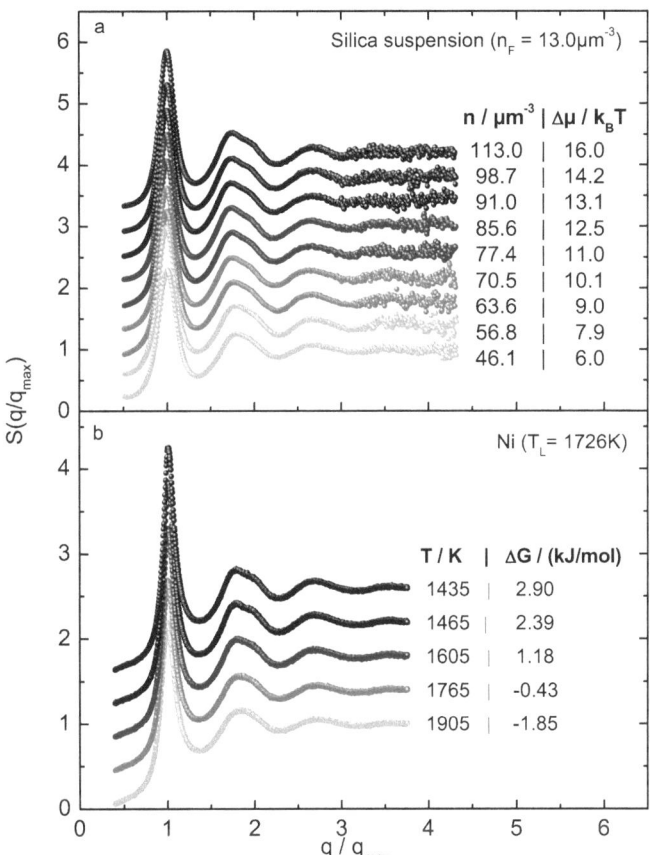

Figure 4.14: Structure factor $S(q/q_{\max})$ as a function of normalized scattering vector $q/q_{\max}$, measured on the melt phase of a colloidal suspension at maximum interaction by USAXS (a) and measured on liquid Ni by neutron scattering (b). For both physically different systems the same behavior is observed. A shoulder appears in the second diffraction peak, which systematically becomes more pronounced with increasing metastability, i.e. increasing chemical potential difference for colloids, $\Delta\mu$, and increasing driving force, ΔG, for metals.

	Ni	Si84
	icosahedral / dodecahedral	icosahedral / dodecahedral
r_0/nm	0.238 / 0.242	315 / 319
$\delta r_0^2 \ / \ nm^2$	$0.29 \cdot 10^{-3} / 0.26 \cdot 10^{-3}$	487 / 435
c / %	99 / 95	99 / 99

Table 4.4: Fit parameters of the simulation procedure of Simonet for the silica system Si84 at $n = 46.1 \mu m^{-3}$ in comparison to a Ni melt at $T = 1435K$ assuming an icosahedral and dodecahedral short-range order.

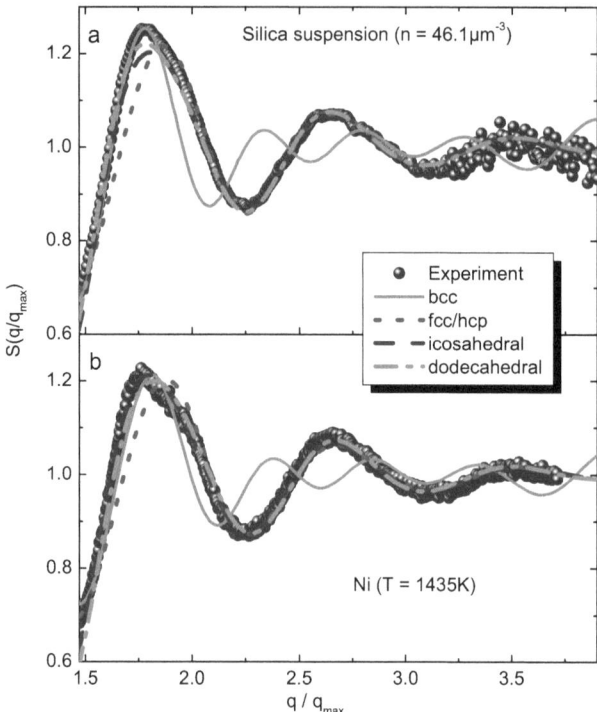

Figure 4.15: Structure factors $S(q/q_{max})$ of a colloidal melt with n $= 46.1 \mu m^{-3}$ (a) and a Ni melt with $T = 1435K$ (b). Measured data are represented by symbols and results of simulations assuming short-range order with different symmetries prevailing in the melt: bcc (dash dotted), fcc and hcp (dotted), icosahedral (dashed) and dodecahedral (solid).

within these clusters. The distance of the central atom to the atoms of the surrounding shell and the nearest neighbor distance within the shell are the same in fcc or hcp clusters but differ of about 5% in icosahedral clusters. The contributions from both these two different intra cluster distances result in a total simulation curve including the asymmetric shape of large oscillations at about $q/q_{max} = 1.8$. The asymmetry in $S(q/q_{max})$ can be therefore seen as an indication of the preference of icosahedral clusters [50, 51].

The comparison of structure factors $S(q/q_{max})$ of a colloidal melt and of molten Ni yielded strikingly similar results if scaled with the scattering vector corresponding to the nearest neighbor distance. In particular for both physically different systems an asymmetric shoulder in the second maximum of $S(q/q_{max})$ was observed which became more pronounced with increasing deviations from equilibrium. Hence, the charged silica system of repulsive interactions is found suitable to model LJ like interactions in metallic systems. Moreover, the experimental data were analyzed by an approach assuming isolated structural units in the liquid [154]. The simulations demonstrate that the assumption of bcc, fcc or hcp structural units does not describe the experimental results for undercooled metallic melts and for the metastable colloidal liquid. The agreement between simulation and experiments becomes near quantitative when icosahedral or dodecahedral clusters are assumed.

Both interaction potentials, the LJ like potential of liquid metals and the Debye-Hückel potential of charged silica spheres, respectively, have soft repulsive contributions and both physically different systems prefer an icosahedral short-range order in the undercooled state. Conversely, hard-sphere systems show no soft but a hard repulsive term. Given this striking similarity in the $S(q/q_{max})$ of metals and charged colloids despite the different type of interactions in both systems, the slightly soft repulsive part of both the LJ like and the Debye-Hückel potential is supposed to form a precondition for the formation of the short-range order in a melt. The finding that the repulsive term of the interaction potential is decisive for the formation of the short-range order in liquids is in full agreement with previous theoretical investigations [66, 159].

In pure metals a short-range order of five-fold symmetry is also present above the melting point. These results are shown in Fig. 4.14 (b) for a Ni melt where structure factors determined below and above the melting temperature are presented. In the stable fluid state at $T = 1765K$ and $T = 1905K$ an asymmetry in the second oscillation of $S(q/q_{\max})$ is observed even if less pronounced. It becomes continuously more pronounced with increasing deviations from the equilibrium state. For colloids, the analogon state for a stable metallic fluid above the melting temperature was discussed in sec. 4.3.3, where the short-range order was investigated near the solid-fluid phase boundary. With increasing NaOH concentration and decreasing strength of interaction due to the screening effect, the asymmetry becomes less pronounced until it disappears completely at $c_{\mathrm{NaOH}} = 3.80 mmol/l$ in the fluid phase. Through the addition of screening electrolyte, the interaction character is changed continuously from soft to nearly hard sphere like behavior. In simulations using hard sphere potentials, no icosahedral short-range is observable. The experimental results obtained for the short-range order in charged sphere colloids near the solid-fluid phase boundary underline the assumption that the soft repulsive interaction is supposed to be responsible for the formation of icosahedral short-range order.

As already mentioned in the previous section, the fundamental difference in phase transitions of metals and colloids lies in the interaction potential. In metals the interaction potentials stay the same for the analyzed system even if the temperature is varied. When increasing the temperature at constant pressure the entropic term of the Gibbs free energy at a phase transition is changed and structures with a low degree of order are favored. Therefore a state like the fluid state is stable at high temperatures. For the liquid phase an increase of T will lead to a less pronounced short-range order, but due to the unchanged interaction potentials the general type of short-range order will not change as experimentally observed in the melt phase of Ni (see 4.14 (b)). For the charge stabilized colloidal system however, the interaction potential is changing by variation of the concentration of screening electrolyte and the temperature always stays constant even at solid-liquid phase transitions. The corresponding change of the enthalpic terms on the one hand may influence the type of short-range order of the fluid phase and it also determines the phase stability on the other hand. Therefore, a change of icosahedral type of short-range order to an fcc structure is observed at high concentrations of the screening electrolyte.

The behavior of the short-range order of the investigated silica suspension also confirms assumptions hypothesized by Frank [2]. In 1952, he postulated that the short-range order in undercooled metallic melts is based upon icosahedral aggregates independent of the structure of the corresponding solid phases. Frank showed that the energy of an icosahedral aggregate of 13 Lennard-Jones atoms is 8% lower than that of crystallographic clusters with the same number of atoms (e.g. aggregates of fcc or hcp structure). The energy calculations for an icosahedral cluster include no explicit assumptions on the degree of undercooling or metastability. Frank's hypothesis does not exclude the preference of icosahedral type of short-range order in stable liquids above the melting point.

Therefore the present work identified a colloidal system consisting of charged silica particles that behaves as a metal concerning the formation of short-range order in the metastable liquid state. In detail an icosahedral short-range order is observed, as it is the case in most pure metallic melts. This finding is of general importance since short-range order essentially governs the formation of solid-liquid interfaces and crystal nucleation in undercooled melts. Owing to the fact that an icosahedral short-range order is incompatible with the translational invariance of crystalline phases, for metallic systems a comparatively high energy of the interface between an undercooled liquid and a nucleus of a non-complex crystalline phase that is formed in the melt is predicted [50, 160]. Obviously, colloids are proper model systems for metals to study short-range ordering. In the next section their model character concerning the nucleation behavior is discussed in addition to their impact on solid-liquid interfaces within the framework of classical nucleation theory.

4.4 Nucleation of colloidal crystals

The understanding of solidification is a long-standing scientific challenge and is subject of intense research interest. Progress in understanding nucleation phenomena has been made using colloidal systems as model systems. Charged sphere colloidal suspensions are an important model system for the experimental study of crystallization. In contrast to atomic systems they provide a sufficiently long experimental time scale for investigations of nucleation. This is partly due to their large size of their elementary particles (hundreds of nm) and the Brownian nature of their motion. Therefore, time-dependent investigations of the nucleation process are possible in colloids. This fact makes mesoscopic model systems very interesting with respect to metals and allows insights in nucleation kinetics behavior which is not accessible in liquid metals for direct observation. Classical nucleation

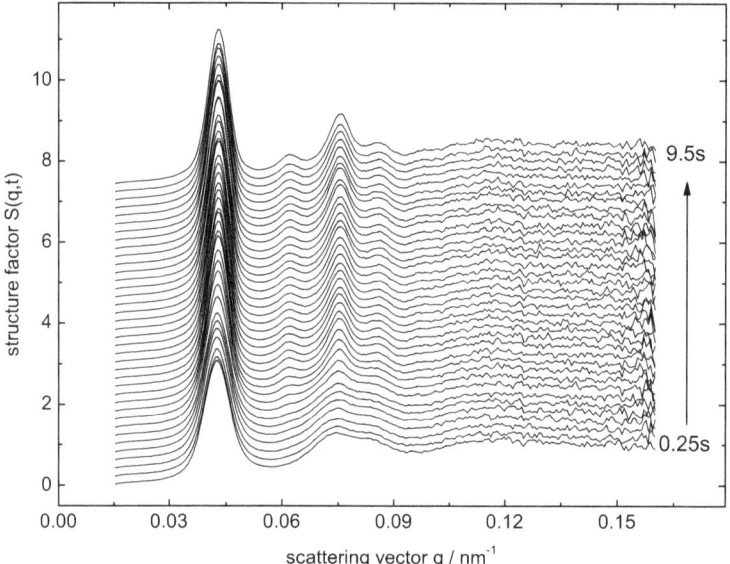

Figure 4.16: Time evolution of the structure factor for the Si77 system at particle number density $n = 224.7 \mu m^{-3}$ and sodium hydroxide concentration of $c_{NaOH} = 2.0 mmol/l$ detected with a time resolution of 0.25s. $S(q,t)$ are shifted for clarity and time increases from bottom to top between 1.25s and 9.5s. The read out time of the detector is on the order of $10^{-3}s$ and therefore neglectable for the time evolution of $S(q,t)$.

theory (CNT) assumes that crystallization is initiated by nucleation at random sites after a delay or induction time depending on the undercooling of the melt. The nucleation process is characterized by the time dependent nucleation rate density $J(t)$, which is defined as the number of nuclei appearing per unit volume and time.

For determination of the nucleation rate density J, it is necessary to observe the evolution of the crystallites from the metastable fluid state into the stable solid state. This process can be analyzed indirectly by scattering methods where the time dependent structure factor $S(q,t)$ is measured. The evolution of $S(q,t)$ is shown in Fig. 4.16 for the silica system Si77 at the particle number density $n = 224.7 \mu m^{-3}$. As long as the system is not completely crystallized the scattered signal includes information of both the crystalline and the fluid phase. Scattering from the remaining fluid in the sample is subtracted by use of the first scan of the crystallizing sample as an approximation to the fluid background $S_f(q) = S_{bg}(q, t = 0)$ following the method of Harland and van Megen [161]. The

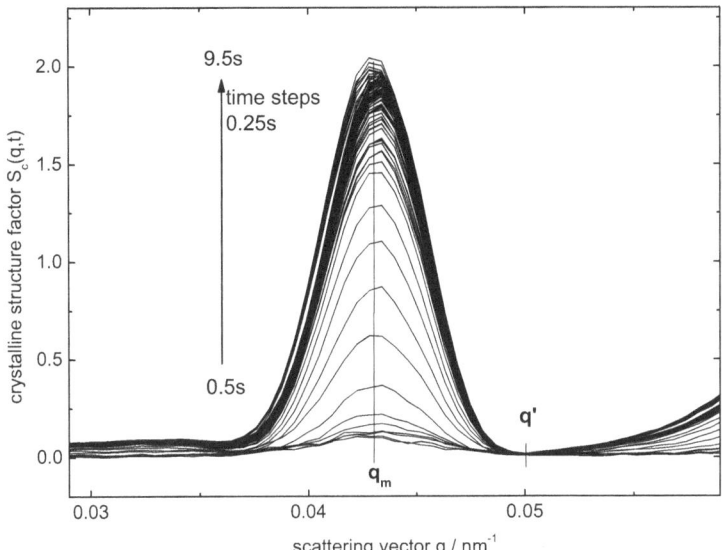

Figure 4.17: The crystalline part of the structure factor $S_c(q,t)$ of the (110) bcc peak for the silica system Si77 at particle number density $n = 224.7\mu m^{-3}$ and sodium hydroxide concentration of $c_{\text{NaOH}}=$ $2.0 mmol/l$ detected with a time resolution of 0.25sec. $S_c(q,t)$ can be determined by subtracting the scaled fluid from the measured structure factor. The integrated intensity is related to the amount of crystals, the peak position q_m to the density in the crystalline phase and the half width to the average size of the crystallites.

background $S_f(q)$ has to be scaled so that the intensity at the minimum q' of the structure factor is equal to the intensity recorded for the following time scans $S(q',t) = \beta(t) \cdot S_f(q')$ with a scale factor $\beta(t)$ chosen at each time step. This scaling assumes that the fluid density and composition retain the same throughout the crystallization. The crystalline part of the structure factor can be determined by subtracting the scaled fluid from the measured structure factor

$$S_c(q,t) = S(q,t) - \beta(t) \cdot S_f(q). \tag{4.9}$$

The resulting crystal structure factor $S_c(q,t)$ is used to determine the properties of the crystalline phase. The integrated intensity is related to the amount of crystalline phase, the peak position q_m to the density in the crystalline phase and the half width to the average size of the crystallites. The definition of the parameters q_m and q' is outlined in Fig. 4.17.

The integrated intensity of the peak $I_{\text{peak}}(t)$ is defined as follows

$$I_{\text{peak}}(t) = 2 \int_{q_m}^{q'} S_c(q,t)\, dq. \tag{4.10}$$

The half width of the peak is given by $\Delta q = 2\left(q_{1/2} - q_m\right)$ with $q_{1/2}$ defined as

$$S_c(q_{1/2}, t) = \frac{S_c(q_m, t) + S_c(q', t)}{2}. \tag{4.11}$$

The half width depends on the crystal size L following the Scherrer equation (3.6) $L = 2\pi K/\Delta q$.

The fraction of the sample which is crystalline, the crystallinity $X(t)$, is proportional to the peak integral and can be calculated from the normalized intensity:

$$X^*(t) = 2c \int_{q_m}^{q'} S_c(q,t) \, dq \qquad (4.12)$$

with the normalization factor c determined from the condition that the final peak intensity is $X^*(t_f) = 1$ representing the fully crystallized sample. Figure 4.19 a) shows a typical result for the crystallinity as a function of time, calculated from the area under the Bragg peak by Eq. (4.12).

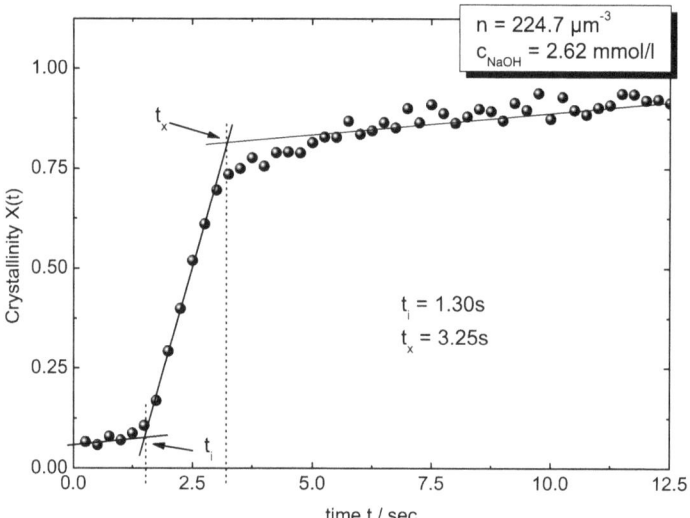

Figure 4.18: Cristallinity X vs. time showing the definition of the induction time t_i and the crossover time t_x for a charged stabilized silica system with particle radius of 77nm and particle number density $n = 224.7 \mu m^{-3}$ at a sodium hydroxide concentration of $c_{NaOH} = 2.62 mmol/l$.

The typical crystallization experiment in charged colloids shows a sigmoidal curve of crystallinity vs. time such as shown in Fig. 4.18. The curve shows initially a sharp increase of the crystallinity which is attributed to nucleation and growth of crystals. Once the sample has reached the equilibrium state, this process decreases. The slow rise at long times is identified as ripening or coarsening process where large crystals grow at the expense of smaller ones. The crystallinity can be qualitatively described by two characteristic time parameters of the crystallization process: the induction time t_i and the crossover time t_x. t_i is obtained by extrapolating a linear fit at the steepest increase in the data to zero crystallinity. t_x is determined by the intersection of linear fits to the initial and long time sections of the experimental data (see Fig. 4.18). The induction and crossover times may be interpreted, respectively, as the time at which nucleation sets in and the time for the completion of crystallization. From the behavior of the crystallinity and the average crystallite size, the time dependent nucleation rate density is derived.

The number density of crystallites is determined from

$$n_{xtal}(t) = \frac{X(t)}{\langle L^3(t) \rangle} = \frac{X(t)}{\alpha \langle L(t) \rangle^3} \qquad (4.13)$$

with the parameter $\alpha \approx 1.25$ relating the average crystal size cubed [162].

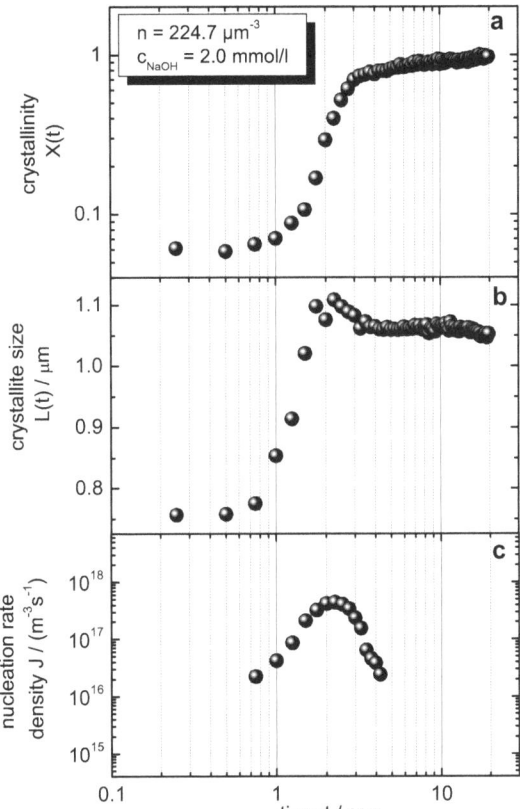

Figure 4.19: Time traces of extracted nucleation parameters from scattering data shown in Fig. 4.16: (a) crystalline volume fraction $X(t)$ or crystallinity, (b) average crystallite size $L(t)$ and (c) nucleation rate densities $J(t)$ exemplarily shown for a charged silica system with particle radius of 77nm and particle number density $n = 224.7 \mu m^{-3}$ at a sodium hydroxide concentration of $c_{\text{NaOH}} = 2.0 mmol/l$.

The nucleation rate density is defined as the rate at which crystals appear in the liquid free volume:

$$J(t) = \frac{1}{(1-X(t))}\frac{d}{dt}\frac{X(t)}{\langle L^3(t) \rangle}. \tag{4.14}$$

This quantity represents the number of critical nuclei which form inside a unit volume of the undercooled liquid. That is why it is necessary to normalize the nucleation rate with the remaining liquid volume, $1-X(t)$, of the sample. Fig. 4.19 shows the time-dependent behavior for a range of parameters extracted from sequence of the structure factors shown in Fig. 4.16: (a) crystallinity X, (b) average crystal size L and (c) nucleation rate densities J.

The behavior of crystallinity X was already discussed in this section and is shown for comparison in Fig. 4.19. The average crystal L grows to a size of about $1.1 \mu m$ and the nucleation rate density first increases from $J = 10^{16} m^{-3} s^{-1}$ by about two orders of magnitude, achieves a maximum value at $J = 8 \cdot 10^{17} m^{-3} s^{-1}$ and decreases again. The decrease of J is attributed to the fact that if no free volume remains available for crystal growth, the nucleation rate density collapses necessarily [82,83].

The introduced parameters are important properties for understanding and describing the nucleation process in colloids as well as in atomic or molecular systems. The time dependence of the nucleation parameters is often not accessible experimentally in both systems, colloids and metals. Colloidal systems, however, show a more sluggish behavior concerning the nucleation and crystallization due to the mesoscopic length scale of the particle sizes and distances. Combined with powerful measurement techniques of suitable time resolutions, time dependent measurements of the nucleation behavior are possible. Due to fast relaxation times, for metals it becomes even more difficult do determine the same quantities time-dependently. This is possible for glass forming alloys only [79] for which the atomic dynamics is very sluggish at temperatures close to the glass transition temperature.

4.4.1 Crystallization kinetics

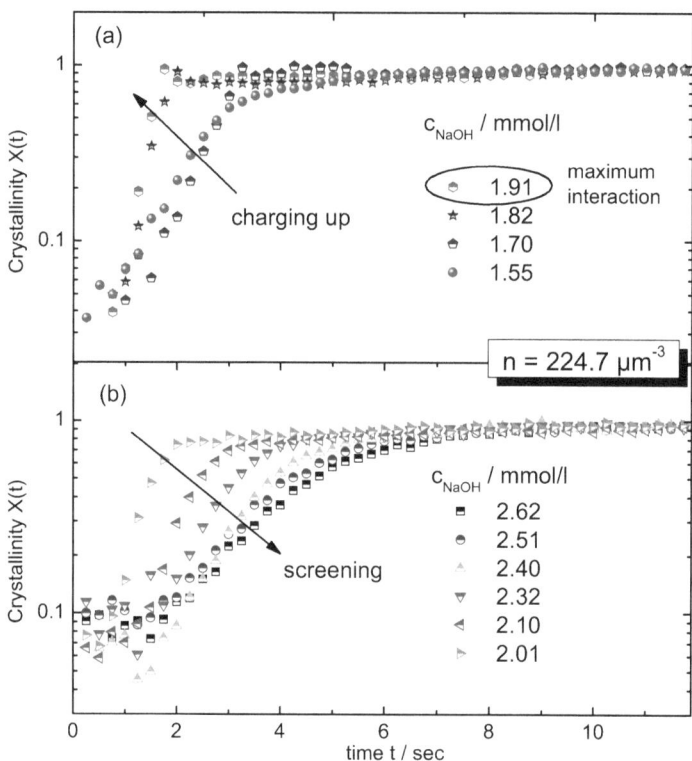

Figure 4.20: Influence of the strength of interaction on the nucleation kinetics. The electrostatic repulsion increases with increasing NaOH concentration up to $c_{NaOH} = 1.9 mmol/l$ for the constant particle number density $n = 224.7 \mu m^{-3}$. At higher concentrations of sodium hydroxide the interaction decreases again due to the excess of counter ions which screen the particles' charge. The curves are separated for a better overview. Figure (a) shows the time-dependent crystallinity for the charging-up process up to $c_{NaOH} = 1.9 mmol/l$ and figure (b) shows the crystallinity behavior influenced by the screening effect.

The nucleation behavior of the silica suspension Si77 strongly depends on the strength of interaction. Fig. 4.20 compares the crystallinity at constant particle number density $n = 224.7 \mu m^{-3}$ in dependence on the sodium hydroxide concentration. The time parameters, t_i and t_x, are dependent on the interaction between the particles which influence the nucleation kinetics. As introduced in sec. 4.2 the electrostatic repulsion increases up to a maximum of the interaction strength at a certain NaOH concentration. At higher concentrations of sodium hydroxide the interaction strength decreases again due to the excess of counter ions which screen the particles' charge. The crystallinity curves are separated in these two stages for a better overview. Figure 4.20 (a) shows the time-dependent crystallinity for the charging-up process up to $c_{NaOH}= 1.9 mmol/l$. Figure 4.20 (b) shows the crystallinity influenced by the screening effect. The most obvious feature of these data is that adding of NaOH causes a faster nucleation process with steeper increase of crystallinity with increasing NaOH concentration up to $c_{NaOH}= 1.9 mmol/l$. Crossing this point by further addition of NaOH leads to a lower slope of crystallinity due to the screening effect.

The kinetic properties of the crystallinity are more obviously reflected in the characterization parameters t_i and t_x. Both values are shown in dependence on c_{NaOH} in Fig. 4.21 for a constant particle number density $n = 224.7 \mu m^{-3}$. There are obvious changes in crystallization kinetics in dependence of the NaOH concentration. The induction time varies slightly between 2.2s and 1.0s. The cross-over time shows a stronger dependence on the strength of interaction. It decreases from 3.8s to 1.7s at $c_{NaOH}= 1.9 mmol/l$ and increases to 7.5s with increasing screening of the particles at $c_{NaOH}= 2.8 mmol/l$. The latter results indicate a retardation of the crystallization process with decreasing electrostatic repulsion or strength of interaction. The shape of both curves shows an initial decrease to a significant minimum in time at $c_{NaOH}= 1.9 mmol/l$ and an increase again at higher sodium hydroxide concentrations. The system displays the fastest kinetics at $c_{NaOH}= 1.9 mmol/l$. This sodium hydroxide concentration is attributed to the state of maximum interaction, where rapid nucleation and growth occur due to the high degree of undercoolability and driving force for crystallization.

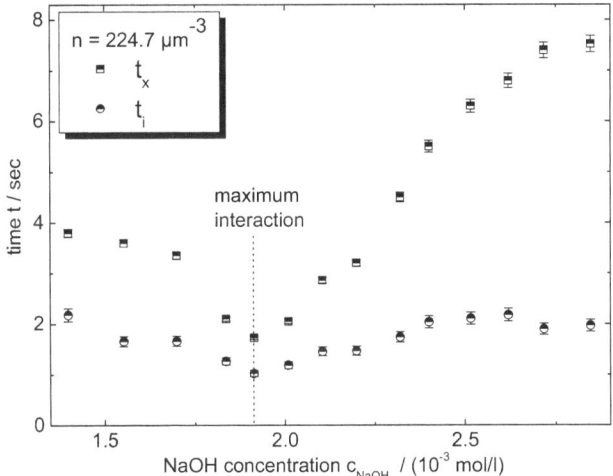

Figure 4.21: The induction time t_i and the cross-over time t_x at constant particle number density $n = 224.7 \mu m^{-3}$ in dependence of sodium hydroxide concentration c_{NaOH}. The shape of both curves shows an initial decrease to a significant minimum in time at $c_{NaOH}= 1.9 mmol/l$ and increases at higher sodium hydroxide concentrations. The sodium hydroxide concentration of $c_{NaOH}= 1.9 mmol/l$ is attributed to the state of maximum interaction, where rapid nucleation and growth occur due to the high degree of undercoolability and driving force for crystallization.

The corresponding time dependent nucleation rate densities resulting from crystallinity data of Fig. 4.20 show a characteristic trend as demonstrated in Fig. 4.22 for the constant particle number density $n = 224.7 \mu m^{-3}$ at various sodium hydroxide concentrations. Each curve shows a significant increase of about one order of magnitude and reaches a maximum when crystallinity is about 80% before decreasing to zero. With increasing interaction ($c_{NaOH}=$ 1.4$mmol/l$ to $c_{NaOH}=$ 1.9$mmol/l$) the magnitude of the maximum nucleation rate densities increases from $J' = 7.0 \cdot 10^{17} m^{-3} s^{-1}$ to $J' = 1.04 \cdot 10^{18} m^{-3} s^{-1}$ while the nucleation time decreases. At maximum interaction, where rapid crystallization is evident, a maximum nucleation rate density of $J' = J_{max} = 1.04 \cdot 10^{18} m^{-3} s^{-1}$ is observed. Crossing the point of maximum interaction by further addition of NaOH ($c_{NaOH}=$ 2.1$mmol/l$ to $c_{NaOH}=$ 2.8$mmol/l$) causes a decrease of the magnitude and an increase of time of the nucleation process reflected in the crossover time t_x. These results clearly show that an accelerated nucleation process takes place at maximum strength of interaction.

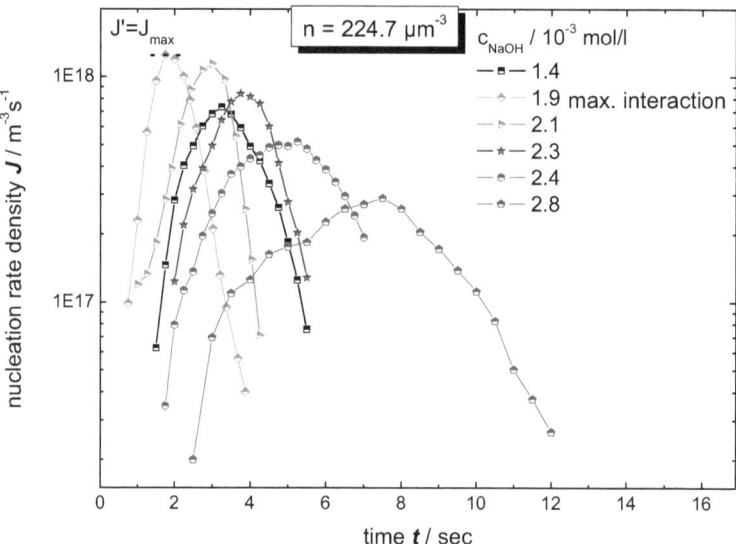

Figure 4.22: Time dependent nucleation rate densities $J(t)$ at constant particle number density $n = 224.7 \mu m^{-3}$ at various sodium hydroxide concentrations. Each curve shows a significant increase of about one order of magnitude and reaches a maximum J' before decreasing to zero. With increasing interaction ($c_{NaOH}=$ 1.4$mmol/l$ to $c_{NaOH}=$ 1.9$mmol/l$) the magnitude of the maximum increases. At maximum interaction for the presented particle number density n, where rapid crystallization is evident, a maximum nucleation rate density of $J' = J_{max} = 1.04 \cdot 10^{18} m^{-3} s^{-1}$ is observed. Crossing the point of maximum interaction by further addition of NaOH ($c_{NaOH}=$ 2.1$mmol/l$ to $c_{NaOH}=$ 2.8$mmol/l$) causes a decrease of the magnitude.

From the time-dependent nucleation rate densities, a constant value can be determined by the maximum point J' on the curve of $J(t)$ as demonstrated in Fig. 4.22. For further evaluation of the nucleation behavior $J'(n) = J_{max}(n)$ only at maximum interaction for each of the investigated particle number densities is taken into account.

On closer inspection of the results in Fig. 4.22 and the characteristic nucleation times, one notes that the induction and crossover times characterizing the time at which nucleation sets in and the time for the completion of crystallization, are of the same order of magnitude which is an indication of transient nucleation effects. Up to now, no transient effects were taken into account in literature for comparable colloidal systems of similar kinetic behavior [5, 161]. The following sections deal with the evaluation of the interfacial free energies using classical nucleation theory. While the evaluation method presented in section 4.4.2 is a simple graphical one where no assumptions on the kinetic prefactor J_0 or the longtime self diffusion coefficient D_S^L are necessary. Sec. 4.4.3 on the other hand presents an evaluation method that considers transient effects using the advanced nucleation theory developed by Collins and Kashchiev [84, 85].

4.4.2 Analysis of the nucleation kinetics within the framework of classical nucleation theory

Homogeneous nucleation is a thermally activated process with competing bulk and surface terms. Nucleation rate densities increase exponentially with undercooling or the difference in the chemical potential between the melt and solid. Thus, nucleation rapidly accelerates with increasing particle number density or strength of interaction.

Despite its simplicity, classical nucleation theory is widely used to describe experimental data on atomic and molecular systems [72, 74, 79], but also for model systems like suspensions of hard or entropically attractive spheres [82, 88, 138, 163–165]. In these systems the interaction potential is independent of the system density. This is not the case for aqueous suspensions of charged stabilized systems. Here any change of particle density will simultaneously alter the surface potential and the screening length. The charged silica system analyzed within this thesis may be considered as representative for a large number of soft condensed matter systems where the interaction between the constituents may be experimentally tuned.

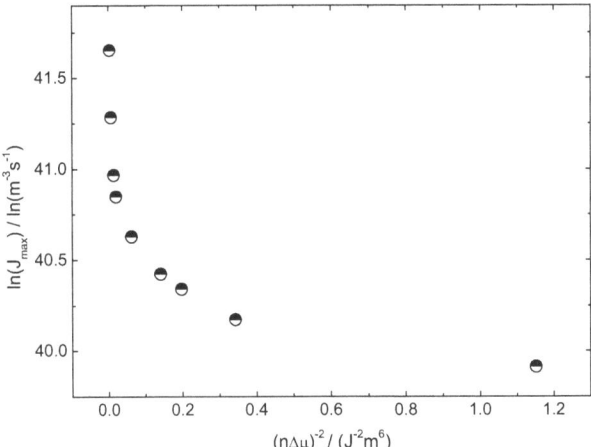

Figure 4.23: Arrhenius plot of the measured nucleation rate densities at maximum interaction versus $(n\Delta\mu)^{-2}$. The plot shows that with increasing undercooling of the metastable melt the logarithm of the nucleation rate density deviates from a linear decay. This behavior can be explained by the concentration dependence of the interfacial free energy of the nuclei. The interfacial free energy γ and the kinetic prefactor J_0 can be determined by the local slope and the intercept with the ordinate at each data point as function of n.

So far, experimental data on charged colloids is still rare [5, 89, 166, 167] and the proof of applicability of classical nucleation theory for density dependent potentials needs further investigations. In this section this question is therefore addressed with a quantitative study of nucleation for charged sphere systems using the prescriptions of CNT adapted to colloidal systems [4, 5].

According to CNT, crystalline clusters are formed in statistically independent nucleation events by stepwise addition of particles from the fluid. The formation is assumed to be thermally activated with a nucleation barrier $\Delta G^* = \frac{16\pi}{3}\frac{\gamma^3}{(n\Delta\mu)^2}$, where γ is the interfacial free energy, n the particle number density and $\Delta\mu$ the chemical potential difference between the melt and solid. For colloids crossing the barrier via diffusion, the rate density of formation of critical nuclei is given as:

$$J = J_0 \exp\left(\frac{-\Delta G^*}{k_B T}\right), \quad (4.15)$$

where J_0 is the kinetic prefactor. While recent computer simulations on mono- and polydisperse hard spheres failed to predict experimental data [168], CNT describes the experimental results well, if the independently determined particle number density dependence of the long-time self-diffusion coefficient D_S^L is taken into account.

In the following the experimentally determined nucleation rate density at maximum interaction in dependence of the particle number density is investigated within the framework of CNT using Eq. (2.46). This equation includes two undefined physical parameters. These are the interfacial free energy γ and the long-time self diffusion coefficient D_S^L.

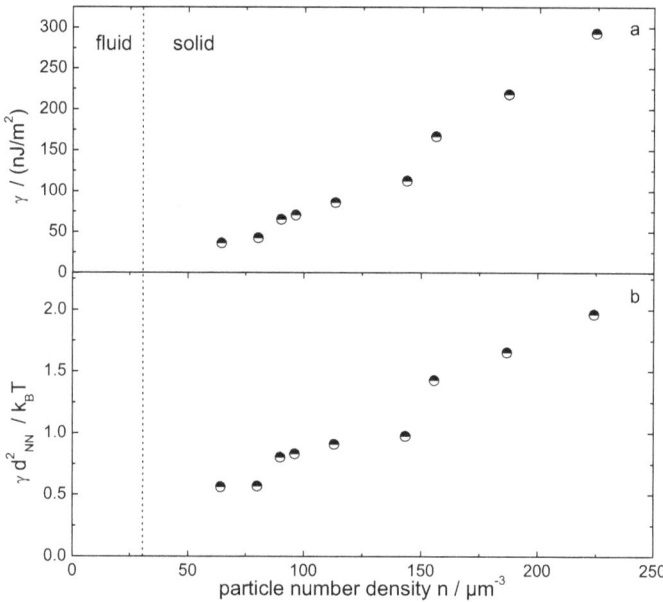

Figure 4.24: Absolute (a) and reduced (b) interfacial energy in dependence of particle number density showing a nearly linear increase.

There are several methods [5] to determine the unknown parameters γ and D_S^L. The simplest one is the graphical method which does not use the exact representation of the kinetic prefactor J_0 according to Eq. (4.15) and thus avoids the need of the exact knowledge of the diffusion coefficient or the interfacial free energy. For this purpose an Arrhenius plot as shown in Fig. 4.23 is used. This plot shows the logarithmic nucleation rate density at maximum interaction $\ln J_{\max}$ against $(n\Delta\mu)^{-2}$ as a measure for the inverse undercooling of the system. $J_{\max}$ is determined from the time-dependent nucleation rate densities by the maximum point J' on the curve of $J(t)$ as demonstrated in Fig. 4.22. Using the logarithm of Eq. (4.15) yields:

$$\ln(J) = \ln(J_0) - \frac{16\pi\gamma^3}{3k_B T}\frac{1}{(n\Delta\mu)^2}. \quad (4.16)$$

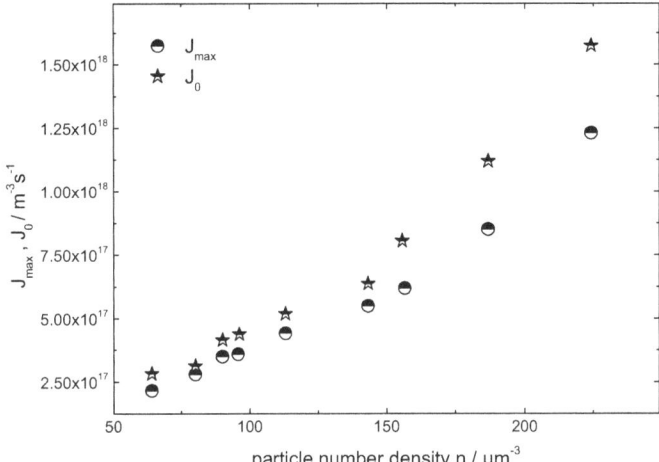

Figure 4.25: Maximum nucleation rate densities J_max and the kinetic prefactor J_0 at maximum interaction as function of the particle number density n.

For a particle density independent interfacial free energy and kinetic prefactor, a straight line with the slope $m = 16\pi\gamma^3/3k_\mathrm{B}T$ according to Eq. (4.16) is expected. Instead, the slope is obviously getting smaller with increasing reduced undercooling.

The non-linear slope in the Arrhenius plot is a clear indication that the interfacial free energy as well as the kinetic prefactor is a function of the particle number density: $m(n)$ and $J_0(n)$. Considering the particle number density dependence of the interfacial free energy and the kinetic prefactor, the local slope of $\ln J_\mathrm{max}$ versus $(n\Delta\mu)^{-2}$ can be calculated using $m(n) = \Delta\ln(J_\mathrm{max})/\Delta(n\Delta\mu)^{-2}$. The concentration dependent interfacial free energy is then given by

$$\gamma(n) = \left[\frac{3m(n)k_\mathrm{B}T}{16\pi}\right]^{1/3}, \tag{4.17}$$

which is represented in Fig. 4.24 (a). The interfacial free energy increases from 30 up to about $300 nJ/m^2$ with increasing n. The kinetic prefactor $J_0(n)$ is determined by the intercept with the ordinate and is shown in Fig. 4.25. In addition also the nucleation rate density at maximum interaction in dependence of the particle number density n is shown. These two data sets have a similar shape and are increasing about one order of magnitude in the investigated range of particle number densities. While at low n a slight difference of the values between $J_\mathrm{max}(n)$ and $J_0(n)$ is observable, the difference increases up to one order of magnitude at high n.

In order to compare the interfacial free energy with atomic systems as well as to other colloidal systems and simulation results, Fig. 4.24 (b) shows the reduced interfacial free energy $\gamma^* = \gamma d_\mathrm{NN}^2$. The reduced interfacial free energy is scaled with the next-nearest neighbor distance d_NN^2 instead of the particle diameter d as usual in hard sphere systems, because charged particles are not close packed. A roughly linear increase of γ^* from $0.5k_\mathrm{B}T$ to $2.0k_\mathrm{B}T$ is observed over the entire accessible particle number density. This result is different to hard sphere systems where a constant value of $\gamma d^2 = 0.55k_\mathrm{B}T$ was obtained [169] which is in good agreement with other experiments on systems of similar spheres [83, 161] and some of theoretical expectation [170]. A direct comparison of experimental and theoretical results of d_NN^2 for various metals and charged stabilized colloids is discussed in sec. 4.4.4.

A further method to obtain the interfacial free energy and the kinetic prefactor, is the direct use of the numerical expression provided by CNT (2.46). This method requires the knowledge of the long-time self diffusion coefficient D_S^L as function of n. With known diffusion coefficient it is possible to solve Eq. (2.46) for γ as the only remaining unknown parameter. The estimate of the diffusion coefficient appears to be quite difficult and remains dependent on assumptions of literature as pointed out in [4–6]. If the reduced interfacial free energy per particle is plotted versus the chemical potential difference, one obtains a plot analogous to the one of Turnbull, recently supplemented by new experimental and theoretical data of metals [79, 171] and colloids [6].

4.4.3 Transient nucleation

The evaluation method presented in the previous section is a simple graphical method which requires no input information about, e.g. the kinetic prefactor or the diffusion coefficient. The diffusion coefficient is often obtained by fitting procedures or theoretical estimates in literature. It is not commonly used to consider transient nucleation effects for colloidal systems of comparable kinetics behavior [5]. The present work presents systematic investigations on the nucleation behavior in charge stabilized colloidal systems where transient effects are considered. There is a discrepancy between the results of both methods in analysis of nucleation behavior observable which requires a critical comparison and discussion.

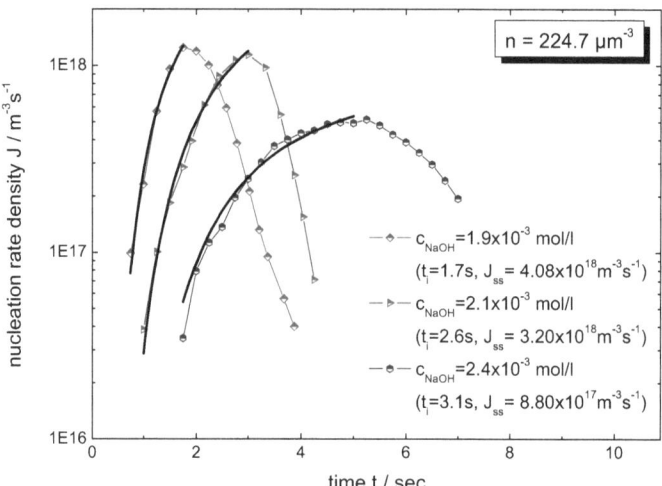

Figure 4.26: Steady state nucleation rate densities J_{ss} and the corresponding induction time t_i obtained following the theory of Kashchiev [85] for the particle concentrations $n = 224.7 \mu m^{-3}$ at various NaOH concentrations.

Fig. 4.18 demonstrates the crystallinity behavior for the Si77 suspension ($n = 224.7 \mu m^{-3}$ and $c_{NaOH} = 2.62 mmol/l$) characterized by the induction or transient time $t_i = 1.30s$. Since the complete nucleation process lasts about 4s until the sample is crystalized and ripening or coarsening effects dominate the behavior of the solid, the transient time cannot be neglected and transient effects have an influence on the nucleation. Therefore, the experimentally determined time-dependent nucleation rate densities as already shown in Fig. 4.22 were further evaluated applying the advanced nucleation theory of Kashchiev [85]. This method allows the determination of the steady state nucleation rate density I_{ss} and the transient time t_i as introduced in sec. 2.6.3. Kashchiev assumed a distribution of relaxation times:

$$J(t) = J_{ss} \left[1 + 2 \sum_{m=1}^{\infty} (-1)^m \exp\left(-\frac{m^2 t}{t_i}\right) \right]. \tag{4.18}$$

where J_{ss} denotes the steady-state nucleation rate and t_i the induction time given by:

$$t_i = \frac{\gamma\, d_{NN}^2\, k_B T}{\frac{3}{8}\left(\frac{4}{3}\right)^{\frac{2}{3}} \pi^{\frac{5}{3}} D n^{\frac{2}{3}} \Delta\mu^2}. \tag{4.19}$$

Three examples of the two parameter fitting procedure for the nucleation data are shown in Fig. 4.26 including the derived transient time t_i and steady state nucleation rate density J_{ss}.

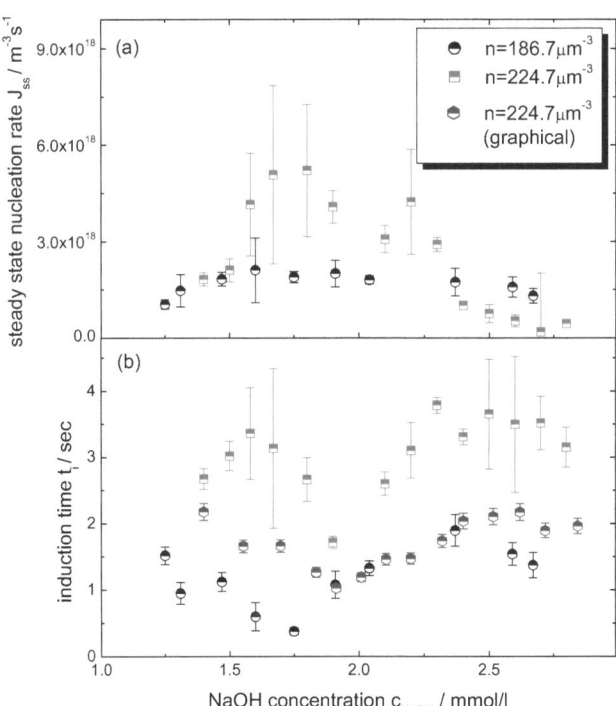

Figure 4.27: Steady state nucleation rate densities J_{ss} and the corresponding induction time t_i obtained following the theory of Kashchiev [85] for the particle concentrations $n = 224.7 \mu m^{-3}$ (squares) and $n = 186.7 \mu m^{-3}$ (circles). The induction time obtained by the graphical evaluation method for $n = 224.7 \mu m^{-3}$ is also plotted for comparison (diamonds).

Fig. 4.27 (a) shows the steady state nucleation rate densities J_{ss} and the corresponding induction time t_i obtained following the theory of Kashchiev [85] for the particle concentrations $n = 224.7 \mu m^{-3}$ and $n = 186.7 \mu m^{-3}$ at various NaOH concentrations. The nucleation rate densities for both n show the same trend to increase with increasing NaOH concentration and to decrease again by approaching the solid-fluid phase boundary at about $c_{NaOH} = 2.8 mmol/l$. The absolute values of J_{ss} are higher for $n = 224.7 \mu m^{-3}$ in comparison to $n = 186.7 \mu m^{-3}$ in the NaOH range between $c_{NaOH} = 1.6 mmol/l$ and $c_{NaOH} = 2.3 mmol/l$. Coincidentally, a higher scatter of J_{ss} is observed in this NaOH range for the higher particle concentration $n = 224.7 \mu m^{-3}$.

The induction time t_i in Fig. 4.27 (b) shows a minimum value of $t_i = 1.7s$ at $c_{NaOH} = 1.9 mmol/l$ for $n = 224.7 \mu m^{-3}$ (squares) and of $t_i = 0.4s$ at $c_{NaOH} = 1.75 mmol/l$ for $n = 186.7 \mu m^{-3}$ (circles). The induction time obtained by the graphical evaluation method for $n = 224.7 \mu m^{-3}$ is also plotted for comparison (diamonds). The latter values reveal a qualitatively similar behavior to the theoretical approach of t_i with a slight difference in quantity. The induction times proposed by the theory of Kashchiev are about one second higher and show a higher scatter in comparison with results obtained by the graphical evaluation method which is discussed in sec. 4.4.2.

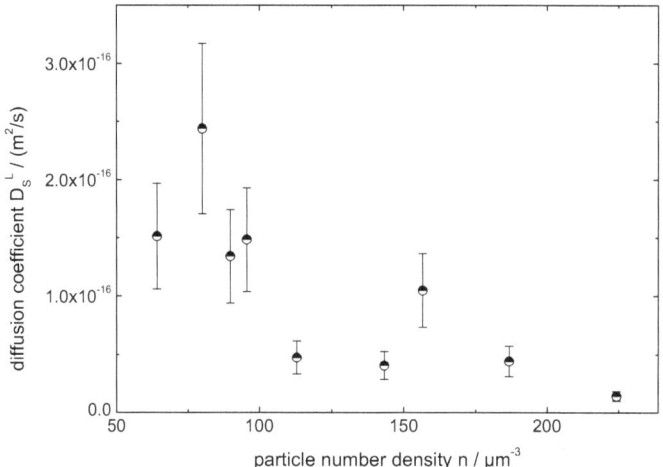

Figure 4.28: Long-time self-diffusion coefficient D_S^L as function of the particle number density at maximum interaction for the Si77 system. Over the entire investigated particle number density n a nearly linear decrease is observed when n is increased.

The steady state nucleation rate density J_{ss} and the induction time t_i are used for the determination of the diffusion coefficient which is an important quantity dealing with crystal nucleation. For charged sphere colloids, the long-time self-diffusion coefficient D_S^L at freezing was determined by forced Rayleigh scattering and from Brownian dynamics simulation without hydrodynamic interactions as follows [172]:

$$D_S^L = 0.1 D_S^S, \quad (4.20)$$

where D_S^S is the measured short-time self-diffusion coefficient [172]. Blees et al. performed extensive NMR measurements of the diffusion coefficient in charged colloids with a 84nm diameter under high salt concentrations over a wide range of particle concentrations up to 15% volume fraction. They found a linear decrease of the self diffusion coefficient with increasing volume fractions [173]:

$$D_S^L = D_0(1 - k\Phi), \quad (4.21)$$

where Φ is the volume fraction, D_0 the Stokes-Einstein diffusion coefficient and $k = 2.85$ a constant value. The Stokes-Einstein diffusion coefficient of an isolated sphere with radius a, suspended in a solvent with viscosity η_0 is defined as:

$$D_0 = \frac{k_B T}{6\pi \eta a}. \quad (4.22)$$

With the viscosity $\eta = 1.002 \cdot 10^{-3} kg/sm$ for water as solvent, the Stokes-Einstein diffusion coefficient yields a value of $D_0 = 5.7 \cdot 10^{-12} m^2/s$ for the Si77 suspension.

Van Blaaderen et al. determined the long-time self diffusion coefficients over a wide range of particle and salt concentrations for a charged sphere system ($a = 122nm$). Their results can be well described by using constants between $k = 3$ and 7 [174]. Wette et al. investigated the nucleation behavior of a charged stabilized polystyrene suspension with a particle diameter of $2a = (68 \pm 3)nm$ and an effective charge of $Z^* = 331$ [5]. These characteristic parameters are similar to those of the

Si77 suspension. They obtained the diffusion coefficients through a fitting procedure by normalizing $D_S^L(n)/D_0 = (a_0 + a_1 n + a_2 n^2)$ to 0.1 at freezing according to Eq. (4.20) using the following approximation for the short-time self diffusion coefficient: $D_S^S = D_0$. Wette et al. observed a nearly linear decrease of the diffusion coefficient with increasing particle number density obtaining a value $k = 4.5$.

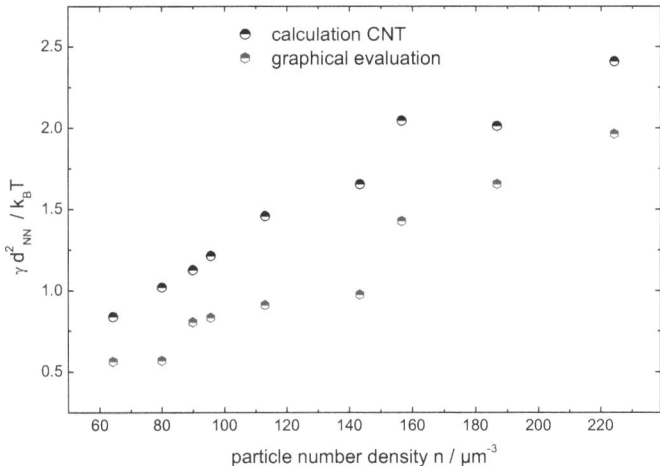

Figure 4.29: Reduced interfacial free energy in dependence of the particle number density. The results obtained by the graphical evaluation method (diamonds) and the direct calculation within the CNT (circles) considering transient effects are shown for comparison.

In contrast to Fig. 4.27 where the long-time self-diffusion coefficient D_S^L is shown for two different particle number densities in dependence of the NaOH concentration, Fig. 4.28 shows the diffusion coefficient as function of the particle number density n at maximum interaction. The diffusion coefficients obtained for the investigated silica suspension of the present work show a nearly linearly decrease in a range from $D_S^L = 1.5 \cdot 10^{-16} m^2/s$ down to $D_S^L = 1.4 \cdot 10^{-17} m^2/s$ with increasing volume fraction. Assuming the Φ-dependence proposed by Blees et al., a constant value of $k = 0.2$ is obtained for the silica suspension Si77. This value is smaller than those obtained from other authors discussed above. Especially the results of Wette et al. differ surprisingly if one considers that the analyzed systems are similar due the particle size and the effective charge. The difference between both systems lies in the polydispersity which is 6% lower for the polystyrene suspension in comparison to the Si77 system (8%). However, polydispersity has a less pronounced influence on the crystal formation in charged stabilized colloids of long-range Coulomb repulsion due to the large inter particle distances in contrast to hard sphere systems and cannot be seen as the main reason for the discrepancy. In addition, Wette et al. did not considered transient effects on the nucleation which can also be taken into consideration as a reason for the difference of the diffusive behavior in both systems. Another reason for the discrepancy in the diffusive behavior may also seen in the investigated regime of volume fractions. Up to now, no experimental data of diffusion coefficients exist for strongly charged colloids with characteristic properties of the Si77 system in the particle concentration regime of up to $224 \mu m^{-3}$. Therefore no direct comparison of experimental results concerning diffusive behavior to literature is available. For comparison only, Wette et al. analyzed the nucleation behavior of similar charge stabilized colloids in a n range between $15 \mu m^{-3}$ and $65 \mu m^{-3}$.

With known diffusion coefficients it is possible to solve Eq. (2.46) for the interfacial free energy γ as the only remaining unknown physical quantity. After obtaining γ, the kinetic prefactor J_0

is accessible. Fig. 4.29 shows the reduced interfacial free energy as a function of the particle number density. The interfacial energy increases nearly linearly from $0.7k_BT$ to $2.4k_BT$ as the particle concentration increases. These values are slightly higher than those obtained by the graphical evaluation method in sec. 4.4.2 which are additionally shown in Fig. 4.29 for a direct comparison. Nevertheless, a qualitatively similar behavior is observed for the results of γd_{NN}^2 obtained by both evaluation methods.

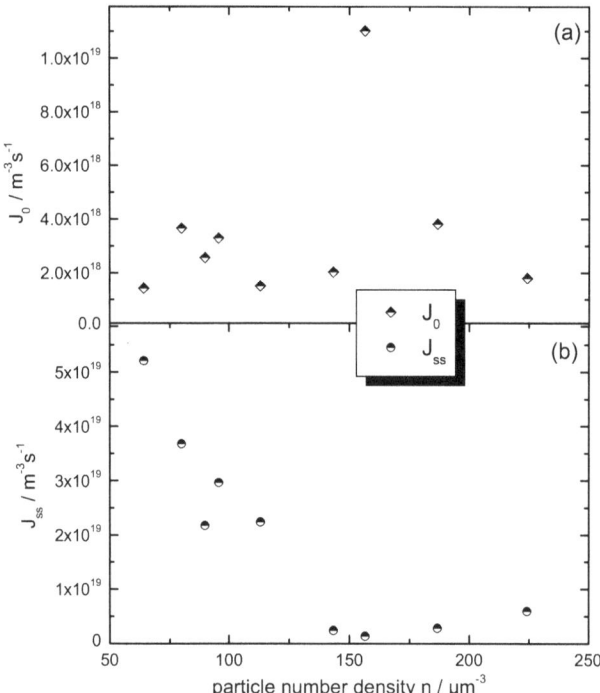

Figure 4.30: Kinetic prefactors (diamonds) and steady state nucleation rate densities (circles) as a function of the particle number density.

The behavior of the kinetic prefactor is shown in Fig. 4.30 (a). It shows a scatter of about one order of magnitude in the investigated particle number range between $J_0 = 1.4 \cdot 10^{18} m^{-3} s^{-1}$ to $J_0 = 1.1 \cdot 10^{19} m^{-3} s^{-1}$. The steady state nucleation rate density shown in Fig. 4.30 (b) decreases from $J_{ss} = 5 \cdot 10^{19} m^{-3} s^{-1}$ to $J_{ss} = 9 \cdot 10^{18} m^{-3} s^{-1}$ as the particle concentration increases. This is different to the behavior of the $J_{max}(n)$ where an increase is observed over the investigated particle number densities. Nevertheless, it stays questionable if both quantities, J_{ss} and J_{max}, respectively, are directly comparable. While J_{ss} is a theoretically derived value representing a constant nucleation rate density, J_{max} results due to the fact of a finite sample cell. If no free volume remains available for crystal growth, the nucleation rate collapses necessarily [82, 83] without achieving the steady state nucleation rate which is necessary for the evaluation of nucleation properties within the framework of the classical nucleation theory.

Summarizing the results, the present study gives a quantitative evaluation of measured nucleation rate densities within the framework of CNT using two different evaluation methods: one simple graphical and a direct calculation method. The previous one requires no exact representation of the kinetic prefactor J_0 and thus avoids the need of the exact knowledge of the diffusion coefficient. The second method directly uses the numerical expression provided by CNT. This method requires the knowledge of the long-time self diffusion coefficient D_S^L as function of n which is accessible through the use of an advanced nucleation theory provided by Kashchiev [85] who considers transient nucleation effects. Regardless of the evaluation method, a qualitative agreement of the results is observable concerning the increasing interfacial free energy γ and the increasing kinetic prefactor J_0 as function of increasing particle number density n. Quantitatively, however, inconsistencies are observable. Using different evaluation methods leads to different results concerning the absolute values of the interfacial free energy and the obtained kinetic prefactors are up to 2 orders of magnitude higher if the graphical evaluation method is used.

4.4.4 Comparison to metals

The interfacial free energy γ is defined as the reversible work required to form a unit area of interface between a crystal and its coexisting fluid. It plays a central role in kinetics of crystal nucleation and growth. Unfortunately, direct experimental investigations of this quantity are difficult and exist only for a handful of materials using grain boundary angles of a fluid groove formed at a solid-liquid interface between two neighboring grains [175]. For most materials, knowledge of the interfacial free energy is obtained indirectly from measurements of nucleation rate densities of undercooled liquids where γ is determined within the framework and approximations of classical nucleation theory. For systems in which directly determined values of γ exist, these are accurate within about 10-20% in comparison to indirect methods. For example, the interfacial energy of bismuth was determined using grain boundary angles to be $61.3 \times 10^{-3} Jm^{-2}$ as compared to the value of $54.4 \times 10^{-3} Jm^{-2}$ obtained indirectly from nucleation rate data [176].

The first systematic investigations of nucleation behavior for a variety of materials were performed by Turnbull in the 1950s [1, 156, 177]. Turnbull reported values of the interfacial energy which were obtained indirectly from nucleation rate data. He used an emulsion technique to reduce heterogeneous nucleation due to impurities and obtained undercooled melts. He dispersed liquid metals in oils or aqueous solutions in the presence of surfactant to stabilize the droplets of 10 to 100 μm in size. If these droplets are sufficiently small, then heterogeneous nucleation sites are confined to a minority of droplets and the probability of any one droplet containing such a site becomes very small. Such impurity free droplets undercool deeply.

In order to compare the results for various systems, Turnbull defined a 'gram-atomic' or molar interfacial free energy of an one atom thick interface containing Avogadro's number of atoms $\gamma^* = \gamma \rho^{-2/3} N_A$. Based on the experimental results, Turnbull proposed an empirical relation between the interfacial free energy γ^* and the latent heat of fusion which corresponds to the chemical potential difference $\Delta \mu$ for colloids:

$$\gamma = C_T \cdot \Delta \mu \cdot n^{\frac{2}{3}}, \tag{4.23}$$

where n is the particle number density and C_T the Turnbull coefficient (or dimensionless solid-liquid interfacial energy). Turnbull obtained a value of $C_T = 0.45$ for most metals especially closed-packed and $C_T = 0.32$ for most nonmetals. Although heterogeneous nucleation could be suppressed to a large degree, Turnbull could not completely avoid the influence of impurities. More advanced methods such as containerless processing have been developed later on and even larger undercoolings could be observed [56] indicating that Turnbull's experiments must have been heterogeneously catalyzed. Today, the values of the reduced interfacial energies obtained by Turnbull around 50 years ago can be assumed as a lower limit for the interfacial energy.

An alternative way to obtain the crystal-melt interfacial energy is to use computer simulations in which the heterogeneous influence on nucleation (e.g. impurities or container walls) can be completely excluded. The interfacial energy stemming from simulations for metals and colloids was compared by Wette within a Turnbull plot [6]. The results are shown in Fig. 4.31. The figure includes simulation results of the interfacial energy calculated by Hoyt [178] for several fcc and bcc metals and those of charged colloidal systems. Hoyt's computer simulations reveal a Turnbull coefficient to be $C_T = 0.55$ for fcc metals and $C_T = 0.29$ for bcc metals. The red diamonds in Fig. 4.31 correspond to the

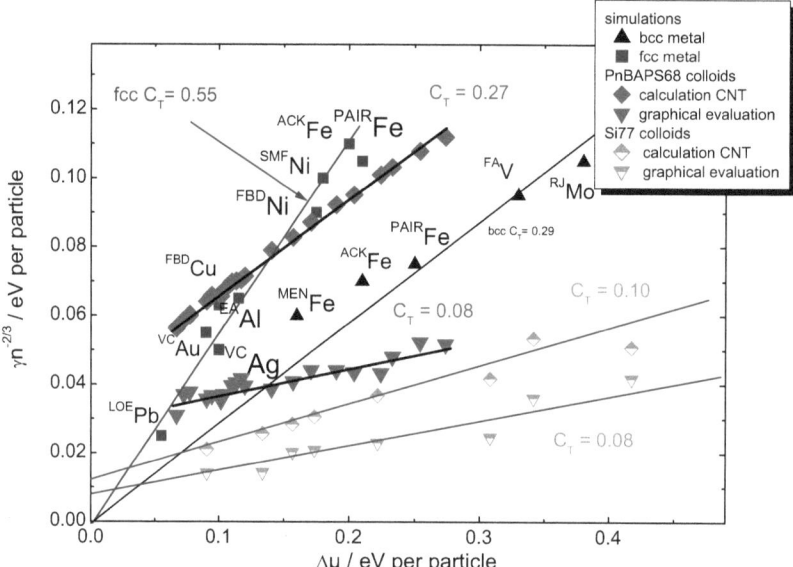

Figure 4.31: Turnbull plot for several metals and colloidal systems. Wette [6] compared the Turnbull coefficient of a charge stabilized colloidal system PnBAPS68 with simulation results for fcc and bcc metals [178]. The labels on each data point refer to the specific form of the embedded atom model (EAM) potential used. In addition the latter results are also compared to the interfacial behavior of the silica system Si77 analyzed within this work. The difference in slopes illustrates the dependence of the Turnbull coefficient C_T on crystal structure for metals and on evaluation methods for colloids.

reduced interfacial free energy of the charge stabilized colloidal system PnBAPS68 [6] which were obtained by two different evaluation procedures. A slope of $C_T = 0.27$ results, if assumptions of the nucleation prefactor are used as an input within the framework of classical nucleation theory and a $C_T = 0.08$ results for the graphical evaluation method which is introduced in sec. 4.4.2 in detail. Equivalent methods were also applied for the silica system Si77 analyzed in the present thesis with the difference that also transient nucleation effects are considered. Here, a similar slope of $C_T = 0.08$ can be observed for the reduced interface energy in dependence on the degree of undercooling derived by the graphical method and a slightly higher Turnbull coefficient of $C_T = 0.10$ is found by the second evaluation method. The latter Turnbull coefficient $C_T = 0.10$ is by the factor 3 smaller compared to $C_T = 0.29$ for the PnBAPS68 suspension. Both values were derived using the direct calculation within the CNT with the difference that the determination of the smaller C_T considers transient effects. The silica system was analyzed at higher particles number densities up to $n = 224.7 \mu m^{-3}$, so that even higher undercoolings were experimentally accessible. Noticeable in the results for both colloidal systems is an offset of the lines which makes no sense from the physical point of view, since the interfacial energy is expected to be zero at infinitesimal degree of undercooling. Up to now, this observation cannot be explained, but it may be an indication for the more complex behavior of the interfacial energy than a simple linear dependence on the degree of undercooling proposed by Turnbull.

The different results between both colloidal systems among themselves evaluated by explicit calculations within the framework of CNT also reflect the difficult interpretation of interfacial energies, which is a long-standing problem in the behavior of atomic systems. The main reason for the discrepancy of the results can be possibly attributed to transient effects which are not considered in the investigations of the PnBAPS68 system. Another reason for the difference in the values of C_T the could be a specific property of colloids, namely the polydispersity, which has no analogon in metals or atomic systems. The polydispersity for the PnBAPS68 system lies in about 2%, while it is slightly higher in the silica system of about 8%. In recent works of Schöpe [83], a significant influence of polydispersity on the nucleation behavior was observed in hard sphere systems. Charged colloids are known to be less sensitive on polydispersity concerning their interaction or structural behavior, but it might play a more significant role in forming a microscopic interface between a metastable fluid and stable solid. Here, more intensive investigations are necessary to clarify the influence of polydispersity in charge colloids concerning their nucleation behavior.

As already mentioned, also for metals it is difficult to compare the interfacial energy obtained by different experimental or theoretical approaches. Nevertheless, the interfacial free energy γ is an important parameter to describe solidification processes within nucleation and growth theories. There exist many models which provide expressions for the energy γ of the interface between the solid and liquid phases independently of the structure. Turnbull recognized that the degree of undercoolability of a metal is strongly dependent on the structure of the melt and the solid state. These observations led to the concept that the failure to undercool some systems was an inherent property that could be explained by the similarity in structure between crystals and their melts. A structural approach to the modeling of the interfacial energy for simple melt-crystal interfaces was developed by Spaepen [179, 180]. The so-called negentropic model of Spaepen is one of the most frequently used models for metals. He suggested to model the interface by randomly dense packing of hard spheres in such a way that the density is maximized to minimize the energy of the interface. The hard sphere approximation leads to a significant simplification of the problem. For metallic systems, however, hard sphere potentials do not provide a very realistic description of the atomic interactions. To compensate this flaw, in the negentropic model further modeling assumptions are made such that a tetrahedral short-range order is preferred and octahedral short-range order is forbidden. The first two construction rules are based upon Frank's prediction of a polytetrahedral short-range order in metallic melts [2]. Spaepen's geometric model assumes that the excess energy of a solid-liquid boundary is negligibly small and he computes an excess configurational entropy from atom packing considerations. He obtains Turnbull coefficients like $C_T = 0.85$ for fcc and $C_T = 0.70$ for bcc forming metals. Tab. 4.5 compiles the experimentally determined Turnbull coefficients for fcc forming solids with $C_T = 0.80$ for Cu [160] and $C_T = 0.67$ for Co [181] and bcc forming solids with $C_T = 0.60$ for Fe [182]. The experimental results of C_T are in a good agreement compared to the predicted Turnbull coefficients by the negentropic model of Spaepen.

Both colloidal systems, PnBAPS68 and Si77, crystallize in a bcc structure and are expected to form a similar short-range order in the undercooled state. Detailed investigations of the short-range order in the metastable state of the Si77 suspension reveal an icosahedral short-range order. Considering these properties within the negentropic model of Spaepen, a Turnbull coefficient of $C_T = 0.70$ is expected for both colloidal systems. This result is not fulfilled for both similar colloidal systems, the Si77 and the PnBAPS68 suspension, respectively, providing different methods to obtain the interfacial free energy.

Using free-energy calculations on small crystalline clusters, Cacciuto et al. estimate the free energy γ for the solid-liquid equimolar interface of a system of hard-sphere colloids [183]. By studying the behavior of a crystallite at coexistence, the dependence of γ on the radius of curvature of the interface was determined. An extrapolation to infinite radius of curvature (flat interface), yields $\gamma(r \to \infty) = 0.616$. Subsequently, he considers the dependence of the interfacial free-energy density on the degree of supersaturation. The simulations suggest that γ associated with the equimolar surface is fairly insensitive to changes in supersaturation [183]. This result obviously does not apply to charge stabilized colloidal systems.

Gránásy et al. developed a phase field theory for the solid-liquid interfacial energy under non-equilibrium conditions of binary systems [184]. The results of undercooling experiments on Ni-Cu melts of different compositions in the full compositional range between 0-100at.%Cu [182] were evalu-

System	Method	stable phase	C_T	Ref.
Si77	graphical	bcc	0.08	this work
Si77	calculation CNT	bcc	0.10	this work
PnBAPS68	graphical	bcc	0.08	[6]
PnBAPS68	calculation CNT	bcc	0.27	[6]
Cu	melt fluxing	fcc	0.80	[160]
Co	melt fluxing	fcc	0.67	[181]
Fe	electromagnetic levitation	bcc	0.60	[182]
$Co_{50}Pd_{50}$	electromagnetic levitation	fcc	0.57	[160]
$Co_{70}Pd_{30}$			0.55	

Table 4.5: Experimental results of the dimensionless solid-liquid interfacial energy (Turnbull coefficient C_T). In case of the results for metals the denoted C_T is seen as a lower limit of the reduced interfacial free energy because heterogeneous nucleation cannot certainly be excluded even if high undercoolings are obtained. In the case of colloidal suspensions heterogeneous contributions can be completely excluded.

ated within this model [184]. For this analysis homogeneous nucleation and a reduced interfacial free energy $C_T = 0.6$ was assumed which is close to the result of MD simulations proposed by Hoyt [185]. A quantitative agreement was achieved with the experimental results in the full compositional range. Gránásy interpreted this behavior as an indication for homogeneous nucleation in the binary Ni-Cu alloy. More recent experimental results, however, revealed even significantly deeper undercoolings on pure Cu using two different experimental undercooling techniques: electromagnetic levitation [186] and melt fluxing [181]. Due to a more efficient reduction of heterogeneous nucleation in the latter experiments, an increase of the maximum undercoolability is achieved ($\Delta T = 310K$ [186] and $\Delta T = 352K$) [181]. This implies that the Turnbull coefficient must be larger than the assumed value of $C_T = 0.6$. Therefore, Hoyt's simulations underestimate the experimental values for the solid-liquid interfacial energy and Spaepen's C_T values predicted by the negentropic model [179, 180] appear more consistent with the experimental results.

To separate between crystallization phenomena induced by homogeneous or heterogeneous nucleation in atomic systems like metals is challenging. The long standing difficulty has occupied many researchers in the last decades. Investigations on the nucleation behavior of undercooled Co_xPd_{100-x} (for $x = 50$ and $x = 70$) melts by application of electromagnetic levitation technique underline the problematic nature of nucleation phenomena [160]. Despite the fact that large relative undercoolings of $\Delta T/T_L \approx 0.21$ were measured, a statistical analysis of the nucleation behavior delivers low nucleation prefactors ($J_0 \approx 10^{23}$) indicating that heterogeneous nucleation prevails in this case. Turnbull coefficients of $C_T = 0.57$ for $x = 50$ and $C_T = 0.55$ for $x = 70$ are observed. Interfacial free energies which are evaluated under the simplified assumptions of homogeneous nucleation are therefore seen as lower limits of the real solid-liquid interfacial energy.

In other approaches, γ can be calculated by molecular dynamics (MD) simulations or in the framework of density functional theory (DFT). Some results of such investigations on the dimensionless solid-liquid interfacial energy C_T are summarized in Table 4.6 (taken from [51]) for different crystal surfaces of phases with fcc or bcc structure. The results of both methods, MD as well as DFT, are strongly dependent on the choice of the interaction potentials, which leads to a broad spread of values for the solid-liquid interfacial energy determined in the different studies. The energy of the interface between crystals with bcc structure and their melt was only determined by Spaepen within the negentropic model ($C_T = 0.70$), by Marr and Gast within DFT calculations $C_T = 0.48$ (for a hard sphere potential) and $C_T = 0.46$ (for a Lennard-Jones potential) and by Hoyt using MD simulations ($C_T = 0.32$ for a Fe-EAM potential and $C_T = 0.29$ for a EAM potential as a mean slope for different metals like Fe, V and Mo). In the work of Broughton and Gilmer [188], Curtin [189] as well as Davidchack and Laird [191], Lennard-Jones and hard-sphere potentials were utilized. The calculations deliver reduced solid-liquid interfacial free energies for fcc crystals between $C_T = 0.36$

Authors	Method	Potential	C_T (fcc)	C_T (bcc)
Spaepen [50]	DPHS	HS	0.85	0.70
McMullen and Oxtoby [187]	DFT	HS	0.87	
Broughton and Gilmer [188]	MD	LJ	0.36	
Curtin [189]	DFT	LJ	0.45	
		HS	0.43	
Marr and Gast [169]	DFT	HS	0.48	0.48
		AS	0.44	0.46
Ohnesorge et al. [190]	DFT	HS	0.18	
		LJ	0.24	
Davidchack and Laird [191]	MD	HS	0.48	
Hoyt et al. [185, 192]	MD	EAM (Ni, Cu, Al, Au, Pb)	0.55	
	MD	EAM (Fe)		0.32
	MD	EAM (Fe, V, Mo)		0.29

Table 4.6: Results if different calculations of the dimensionless solid-liquid interfacial energy (Turnbull coefficient C_T) for crystals with fcc and bcc structure. The table includes abbreviations for different methods: dense packing of hard spheres (DPHS), density functional theory (DFT) and molecular dynamics simulations (MD) using different potentials: hard sphere potential (HS), Lennard-Jones potential (LJ), adhesive sphere potential (AS) and embedded atom method (EAM).

and $C_T = 0.48$. Even smaller values of $C_T = 0.18$ (HS) and $C_T = 0.24$ (LJ) are obtained by Ohnesorge et al. [190] using DFT. It is noticeable that the theoretically calculated values are obviously smaller than the reduced interfacial energy obtained within the negentropic model of Spaepen or by McMullen and Oxtoby [187] which are in the best agreement referring to experimental data for pure metals compiled in Table 4.5. Nevertheless, the comparison between theoretically predicted Turnbull coefficients by different models in Table 4.6 with current experimental ones in Table 4.5 reveals that only the model of Spaepen [50] and the calculations by McMullen and Oxtoby [187] yield Turnbull coefficients that lie within the range of measured values for C_T. Of all discussed approaches, these two models appear to be best suited to describe the solid-liquid interfacial energy of metallic systems forming solid phases with fcc and bcc structures. Therefore a Turnbull coefficient of $C_T = 0.70$ is expected for a colloidal system as a proper model system for nucleation behavior in pure metals which crystallize in a bcc structure. In contrast, smaller Turnbull coefficients of $C_T = 0.10$ and $C_T = 0.08$ are obtained without agreement to experiments of pure metals. In literature, a large variety of theoretical predictions for smaller C_T compared to Spaepen's results exists [169, 185, 188–192]. These values, however, fail to describe the experimental results of metals [160, 160, 181, 182]. As already discussed in the present section, small values of C_T turn out to be heterogeneously influenced in the experiment, if higher undercoolings are achieved due to improvements of experimental techniques [182, 184, 186]. The experimental challenge to separate between crystallization phenomena induced by homogeneous or heterogeneous nucleation in metals has occupied many researchers in the last decades and was realized for a few promising techniques only. For colloidal systems in contrast, the argument of the influence of heterogeneous nucleation effects do not come into question for explaining small C_T. Heterogeneous and homogeneous contributions are also present in colloidal crystals but they can be clearly separated in scattering experiments. The evaluation method for separating heterogeneous nucleation from homogeneous contributions is introduced in sec. 3.4.2 in detail.

The comparison between atomic and colloidal system and the discrepancy concerning the nucleation results can also be seen as an indication of the failure of the simple assumptions and concepts of CNT. The classical nucleation theory assumes statistically independent nucleation events and growth by statistically independent (diffusive) addition of single particles to the nucleus. The probability that a nucleus will change its particle number is independent on the attachment and the dissociation events. CNT takes also no interaction between the appearing nuclei into account. Events that nuclei collide and join together or that the particle flow to a growing nucleus influences the growth of another one are ignored. By describing the nucleus formation the properties of the undercooled melt are also ignored. Density fluctuations in an undercooled melt can cause a sudden appearance of dense locally arrested regions with very slow dynamics which can lead to a crystal nucleus, rather than a stepwise growth by single particles. This was also critically discussed by Oxtoby [193]. It is difficult to conceive that the simple picture used by CNT is still valid in a system with strong and long range interaction. A collective assembly of particles creating a crystal nucleus is supposed to be a possible and more realistic picture. Thus a nucleus appears when particles in the first few coordination shells of the fluid rearrange to a crystal-like order. This could qualitatively explain the results of the present work of unexpected small values of Turnbull coefficients by using the picture of single particle attachment.

Summarizing the results, one can say that the nucleation behavior of colloids can be investigated within the framework of CNT in a similar way as it works for metals or more general speaking atomic systems. The best theoretical description of C_T in metallic systems is given by the negentropic model of Spaepen [179,180] and by the calculations of McMullen and Oxtoby [187]. The negentropic model by Spaepen is the first model that describes the interfacial energy as a function of the structure of both nucleus and interface. He suggested to model the interface by randomly dense packing of hard spheres in such a way that the density is maximized to minimize the energy of the interface. The hard sphere approximation leads to a significant simplification of the problem and is obviously not fulfilled in charged stabilized colloids. Here, crystal phases at low volume fractions and at the presence of the solution medium do not resemble the dense random packing arrangements. This property is supposed to be the main reason for the discrepancy in the results of interfacial energies between colloids and pure metals. Additionally, the question still remains, wether the simple picture and concepts of CNT are valid for the complex behavior of nucleation phenomena for colloids as well as metals, independently of length scales at which nucleation occurs.

Chapter 5
Summary and Outlook

The aim of the present thesis employing charge stabilized colloidal suspensions is to study their model character for pure metals. For this challenging task the main prerequisite is the availability of a well characterized system and the possibility to tune the interaction in a precise and reversible way. The possibilities of tuning the interactions in colloidal charged sphere suspensions have been shown and a large variety of systematic measurements on the crystallization kinetics as well as on properties of melts and resulting solids have been performed including measurements of growth and elasticity. In many cases a great analogy exists between colloidal and metal systems. Under these conditions systematic investigations on the short-range order in and out of the equilibrium state were performed revealing an icosahedral short-range order in the undercooled state over the entire accessible particle concentrations. The icosahedral short-range order becomes more pronounced at increasing deviation from equilibrium state. However, when tuning the interaction of charged colloids from soft repulsive to a nearly hard-sphere like potential, the icosahedral short-range order vanishes and a fcc like order in the fluid state is observed.

Various melts of pure metals, e.g. Fe, Ni, Zr or Cu also show an icosahedral short-range order in the undercooled state. The icosahedral character of metallic melts decreases with decreasing undercooling, but is also present in the state above the melting point. There is a fundamental difference in the phase transition behavior between charge stabilized colloids and metals. In metals the interaction potentials stay the same for the analyzed system even if the temperature is varied. With increasing temperature structures with a low degree of order will be favored. Therefore a state like the fluid state is stable at high temperatures. For the liquid phase an increase of temperature will lead to a less pronounced short-range order, but due to the unchanged interaction potentials the general type of short-range order will not change as experimentally observed. For the investigated colloidal system however, the interaction potential is changing by variation of the concentration of screening electrolyte and the temperature always stays constant even at solid-liquid phase transitions. The corresponding change of the enthalpic terms on the one hand may influence the type of short-range order of the fluid phase, on the other hand it also determines the phase stability as function of the screening electrolyte concentration. The fluid state is achieved continuously by strong screening of charged spheres with electrolytes. The latter state corresponds to a hard sphere like behavior where no soft repulsive interactions are present. The experimental results obtained for the short-range order in charged sphere colloids in comparison to pure metals underline the assumption that the soft repulsive interaction is supposed to be responsible for the formation of icosahedral short-range order.

In addition to investigations of the short-range order, further priority of the present work was dedicated to the question if charged colloids are suitable model systems for studying nucleation phenomena in metals. Time resolved measurements were employed using USAXS technique at HASYLAB (DESY). This allowed a quantitative access to the nucleation kinetics and compiles a comprehensive data base covering several important situations. Both the reaction controlled growth after heterogeneous nucleation and the activated process of homogeneous nucleation could be monitored with high accuracy and the data could well be parameterized using classical phenomenological theories. Due to the relatively convenient experimental accessibility of colloidal melts the data are

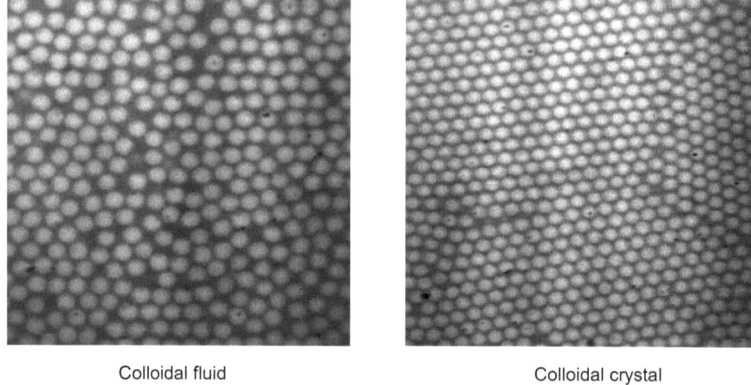

Colloidal fluid   Colloidal crystal

Figure 5.1: Confocal microscopy images of a colloidal fluid (a) and a solid (b) where the particles ($2\mu m$ in diameter). Short-range and long-range arrangements of colloidal particles are resolved in real-space. In the second case also a grain boundary is visible. Images are presented with kind permission of P. Wette.

of high statistical accuracy, they are free of heterogeneous nucleation contributions, the interaction potential is well known and even the melt structure is accessible. The present work also considered the influence of transient effects on nucleation behavior and allowed a verification of its influence on nucleation phenomena in colloidal systems. Together this allowed a critical assessment of classical nucleation theory. The CNT with the key parameters of interfacial energy and kinetic prefactor indeed provides a useful parameterization, but still a large number of challenges remain, e.g. to describe the relation of nucleation rate densities and melt structure beyond qualitative analogies, or to address the case of heterogeneous nucleation which is so important in metal systems.

In order to compare the results of interfacial free energy for various systems, Turnbull defined a 'gram-atomic' or molar interfacial free energy of an one atom thick interface [1,156,177]. Based on the experimental results, Turnbull proposed an empirical relation between the molar interfacial free energy γ^* and the latent heat of fusion which corresponds to the chemical potential difference $\Delta\mu$ for colloids: $\gamma^* = C_T \Delta\mu n^{\frac{2}{3}}$, where C_T is the Turnbull coefficient or dimensionless solid-liquid interfacial energy. Although heterogeneous nucleation could be suppressed to a large degree, Turnbull could not completely avoid the influence of impurities. More advanced methods such as containerless processing have been developed later on and even larger undercoolings could be observed [56] indicating that heterogeneous nucleation governs the metallic melts in Turnbull's experiments. Consequently, the values of the reduced interfacial energies obtained by Turnbull around 50 years ago can be assumed as a lower limit for the interfacial energy. Actual experimental results reveal higher Turnbull coefficients for metals. The best theoretical description of C_T in metallic systems is given by the negentropic model of Spaepen [179,180] and by the calculations of McMullen and Oxtoby [187]. The negentropic model by Spaepen is the first model that describes the interfacial energy as a function of the structure of both nucleus and interface. He suggested to model the interface by randomly dense packing of hard spheres in such a way that the density is maximized to minimize the energy of the interface. Moreover a polytetrahedral short-range order is assumed to prevail in the solid-liquid interface. The negentropic model predicts a Turnbull coefficient of $C_T = 0.7$ for metals forming a bcc structure in the solid. The presented evaluation methods in this work demonstrate essentially smaller Turnbull coefficients of $C_T = 0.1$ and $C_T = 0.08$, respectively, for the colloidal system Si77. For charge stabilized colloidal systems studied in the this work the crystal phases at low volume fractions and at the presence of the solution medium between the particles cannot be considered as densely packed systems. This difference is supposed to be the main reason for the discrepancy in the results of interfacial energies between colloids and metals.

It often has been pointed out that a microscopic theory of crystallization is still missing. Here colloidal systems may play an important role. Detailed investigations of phase transitions in colloids [138] are possible even in real-space [69]. Using confocal microscopy, spatial resolutions in the micron regime are possible to observe the formation of short-range order, nucleation and solidification of colloids in real-space as demonstrated in Fig. 5.1. These unique experimental advantages in combination with tunable interactions mark out colloids as excellent model systems for atomic materials. Microscopic experiments with single particle resolution have just begun to reveal the details of the nucleation mechanism. If such experiments are cross checked with macroscopic ones, like those presented here and combined with careful computer simulation, a full description of nucleation processes in simple systems could be within reach soon.

Bibliography

[1] Turnbull, D. *Thermodynamic driving force for nucleation and growth*. J. Appl. Phys **21**, 1022–1029 (1950).

[2] Frank, F. C. *Supercooling of Liquids*. Proceedings of the Royal Society of London. Series A, Mathematical and Physical Sciences (1934-1990) **215**, 43–46 (1952).

[3] Wette, P., Klassen, I., Holland-Moritz, D., Palberg, T. and Herlach, D. M. *Colloids as model systems for liquid undercooled metals*. Phys. Rev. E **79**, 010501 (2009).

[4] Wette, P., Schöpe, H.-J., Liu, J. and Palberg, T. *Solidification in model systems of spherical particles with density-dependent interactions*. Europhys. Lett. **64**, 124–130 (2003).

[5] Wette, P. and Schöpe, H. J. *Nucleation kinetics in deionized charged colloidal model systems: A quantitative study by means of classical nucleation theory*. Phys. Rev. E **75**, 051405 (2007).

[6] Wette, P. *Eigenschaftskorrelationen in kolloidalen Festkörpern und Fluiden aus optischen Experimenten*. Ph.D. thesis, Johannes Gutenberg-Universität Mainz (2006).

[7] Yethiraj, A. *Tunable colloids: control of colloidal phase transitions with tunable interactions*. Soft Matter **3**, 1099–1115 (2007).

[8] Lin, M. Y. et al. *Universality in colloid aggregation*. Nature **339**, 360–362 (1989).

[9] Hamaker, H. C. *The London-van der Waals attraction between spherical particles*. Physica **4**, 1058–1072 (1937).

[10] Israelachvili, J. N. *Intermolecular and surface forces* (Academic Press, New York, 1992).

[11] Hansen, J. P. and McDonald, I. R. *Theory of simple liquids* (Academic Press, 2006).

[12] Rudhardt, D., Bechinger, C. and Leiderer, P. *Direct Measurement of Depletion Potentials in Mixtures of Colloids and Nonionic Polymers*. Phys. Rev. Lett. **81**, 1330–1333 (1998).

[13] Sedgwick, H., Egelhaaf, S. U. and Poon, W. C. K. *Clusters and gels in systems of sticky particles*. J. Phys. Condens. Matter **16**, 4913–4922 (2004).

[14] Wood, W. W. and Jacobson, J. D. *Preliminary Results from a Recalculation of the Monte Carlo Equation of State of Hard Spheres*. J. Chem. Phys. **27**, 1207–1208 (1957).

[15] Alder, B. J. and Wainwright, T. E. *Phase Transition for a Hard Sphere System*. J. Chem. Phys. **27**, 1208–1209 (1957).

[16] Pusey, P. N. and Van Megen, W. *Phase behavior of concentrated suspensions of nearly hard colloidal spheres*. Nature **320**, 340–342 (1986).

[17] Pusey, P. N., Levesque, D. and Zinn-Justin, J. *Liquids, Freezing, and Glass Transition (Proc. Les Houches Summer School of Theoretical Physics 1989, Part II)* (ed. Hansen, J. P., Amsterdam, North-Holland, 1991).

[18] Stern, O. *Zur Theorie der elektrolytischen Doppelschicht*. Zeitschrift für Elektrochemie **30**, 508–516 (1924).

[19] Helmholtz, H. *Studien über elektrische Grenzschichten bei galvanischer Spannung und der durch Wasserströmung erzeugten Potentialdifferenz*. Annalen der Physik und Chemie **7**, 337–382 (1879).

[20] Gouy, M. *Sur la constitution de la charge électrique à la surface d'un électrolyte*. J. Phys. Theor. Appl. **9**, 457–468 (1910).

[21] Chapman, D. L. *A contribution to the theory of electrocapillarity*. Phil. Mag **25**, 475–481 (1913).

[22] Lobaskin, V., Dünweg, B., Medebach, M., Palberg, T. and Holm, C. *Electrophoresis of Colloidal Dispersions in the Low-Salt Regime*. Phys. Rev. Lett. **98**, 176105 (2007).

[23] Debye, P. and Hückel, E. *The interionic attraction theory of deviations from ideal behavior in solution*. Z. Phys **24**, 185 (1923).

[24] Debye, P. and Hückel, E. *On the Theory of Electrolytes. III: Osmotic Equation of State and Activity of Diluted Strong Electrolytes*. Z. Physik. **25**, 97 (1924).

[25] Alexander, S. et. al. *Charge renormalization, osmotic pressure, and bulk modulus of colloidal crystals: Theory*. J. Chem. Phys. **80**, 5776–5781 (1984).

[26] Robbins, M. O., Kremer, K. and Grest, G. S. *Phase diagram and dynamics of Yukawa systems*. J. Chem. Phys. **88**, 3286–3312 (1988).

[27] Palberg, T., Mönch, W., Bitzer, F., Piazza, R. and Bellini, T. *Freezing Transition for Colloids with Adjustable Charge: A Test of Charge Renormalization*. Phys. Rev. Lett. **74**, 4555–4558 (1995).

[28] van Roij, R. *Attraction or repulsion between charged colloids? A connection with Debye-Huckel theory*. J. Phys. Condens. Matter **12**, 263–268 (2000).

[29] Derjaguin, B. and Landau, L. D. *A theory of the stability of strongly charged lyophobic sols and the coalescence of strongly charged particles in electrolytic solution*. Acta Phys.-Chim. USSR **14**, 633–662 (1941).

[30] Verwey, E. J. W. and Overbeek, J. *(1948) Theory of the stability of lyophobic colloids*. Elsevier, Amsterdam **1**, 50 (1948).

[31] Crocker, J. C. and Grier, D. G. *When like charges attract: The effects of geometrical confinement on long-range colloidal interactions*. Phys. Rev. Lett. **77**, 1897–1900 (1996).

[32] van Roij, R. and Hansen, J. *Van der Waals-like instability in suspensions of mutually repelling charged colloids*. Phys. Rev. Lett. **79**, 3082–3085 (1997).

[33] van Roij, R., Dijkstra, M. and Hansen, J.-P. *Phase diagram of charge-stabilized colloidal suspensions: van der Waals instability without attractive forces*. Phys. Rev. E **59**, 2010–2025 (1999).

[34] Evers, M., Schöpe, H. J., Palberg, T., Dingenouts, N. and Ballauff, M. *Residual order in amorphous dry films of polymer latices: indications of an influence of particle interaction*. Journal of Non-Crystalline Solids **307-310**, 579 – 583 (2002).

[35] Schöpe, H. J. and Palberg, T. *Frustration of structural fluctuations upon equilibration of shear melts*. Journal of Non-Crystalline Solids **307-310**, 613–622 (2002).

[36] Sirota, E. B. et al. *Complete phase diagram of a charged colloidal system: A synchrotron x-ray scattering study.* Phys. Rev. Lett. **62**, 1524–1527 (1989).

[37] Monovoukas, Y. and Gast, A. P. *The experimental phase diagram of charged colloidal suspensions.* J. Coll. Interf. Sci. **128**, 533–548 (1989).

[38] Jackson, J. D. *Classical Electrodynamics* (Wiley, New York, 1975).

[39] Dhont, J. K. G. *An Introduction to Dynamics of Colloids* (Elsevier New York, 1996).

[40] Berne, B. J., E, R. and Pecora, R. *Dynamic Light Scattering* (1990).

[41] Nägele, G. *On the dynamics and structure of charge-stabilized suspensions.* Physics reports **272**, 215–372 (1996).

[42] Kerker, M. *The scattering of light and other electromagnetic radiation* (Academic Press, New York, 1967).

[43] Würth, M., Schwarz, J., Culis, F., Leiderer, P. and Palberg, T. *Growth kinetics of body centered cubic colloidal crystals.* Phys. Rev. E **52**, 6415–6423 (1995).

[44] Schöpe, H. J., Fontecha, A. B., Konig, H., Hueso, J. M. and Biehl, R. *Fast microscopic method for large scale determination of structure, morphology, and quality of thin colloidal crystals.* Langmuir **22**, 1828–1838 (2006).

[45] Guinier, A., Fournet, G., Walker, C. B. and Yudowitch, K. L. *Small-angle scattering of X-rays* (Wiley New York, 1955).

[46] Feigin, L. A., Svergun, D. I. and Taylor, G. W. *Structure analysis by small-angle X-ray and neutron scattering* (Plenum Press New York, 1987).

[47] Tanford, C. *Physical Chemistry of Macromolecules* (John Wiley & Sons, Inc, 1961).

[48] Chu, B. and Liu, T. *Characterization of nanoparticles by scattering techniques.* Journal of Nanoparticle Research **2**, 29–41 (2000).

[49] Glatter, O. and Kratky, O. *Small angle X-ray scattering* (Academic Press London, 1982).

[50] Nelson, D. R. and Spaepen, F. *Polytetrahedral order in condensed matter.* Solid State Physics **42**, 1–90 (1989).

[51] Herlach, D. M., Galenko, P. K. and Holland-Moritz, D. *Metastable solids from undercooled melts* (Pergamon Materials Series, ed. by R. Cahn, 2007).

[52] Hamilton, E. and Cairns, H. *The Collected Dialogues* (Plato., 1961).

[53] Janot, C. *Quasicrystals: a primer* (Oxford University Press, USA, 1997).

[54] Drehman, A. J. and Turnbull, D. *Solidification behavior of undercooled $Pd_{83}Si_{17}$ and $Pd_{82}Si_{18}$ liquid droplets.* Scripta Metallurgica **15**, 543–548 (1981).

[55] Bardenheuer, P. and Bleckmann, R. *Zur Frage des Erstarrens von Stahl in Gussblöcken.* Naturwissenschaften **29**, 550–553 (1941).

[56] Herlach, D. M., Cochrane, R. F., Egry, I., Fecht, H. J. and Greer, A. L. *Containerless processing in the study of metallic melts and their solidification.* International Materials Reviews **38**, 273–347 (1993).

[57] Reichert, H. et al. *Observation of five-fold local symmetry in liquid lead.* Nature **408**, 839–841 (2000).

[58] Di Cicco, A., Trapananti, A., Faggioni, S. and Filipponi, A. *Is there icosahedral ordering in liquid and undercooled metals?* Phys. Rev. Lett. **91**, 135505 (2003).

[59] Schenk, T., Holland-Moritz, D., Simonet, V., Bellissent, R. and Herlach, D. M. *Icosahedral Short-Range Order in Deeply Undercooled Metallic Melts.* Phys. Rev. Lett. **89**, 075507 (2002).

[60] Kelton, K. F. et al. *First X-Ray Scattering Studies on Electrostatically Levitated Metallic Liquids: Demonstrated Influence of Local Icosahedral Order on the Nucleation Barrier.* Phys. Rev. Lett. **90**, 195504 (2003).

[61] Muhlbach, J., Sattler, K., Pfau, P. and Recknagel, E. *Evidence for Magic Numbers of Free Lead-Clusters.* Physics Letters A **87**, 415–417 (1982).

[62] Steinhardt, P. J., Nelson, D. R. and Ronchetti, M. *Bond-orientational order in liquids and glasses.* Phys. Rev. B **28**, 784–805 (1983).

[63] Steinhardt, P. J., Nelson, D. R. and Ronchetti, M. *Icosahedral Bond Orientational Order in Supercooled Liquids.* Phys. Rev. Lett. **47**, 1297–1300 (1981).

[64] Sachdev, S. and Nelson, D. R. *Order in metallic glasses and icosahedral crystals.* Phys. Rev. B **32**, 4592–4606 (1985).

[65] Yonezawa, F. *Glass transition and relaxation of disordered structures.* Solid State Physics **45**, 179–254 (1991).

[66] Jónsson, H. and Andersen, H. C. *Icosahedral Ordering in the Lennard-Jones Liquid and Glass.* Phys. Rev. Lett. **60**, 2295–2298 (1988).

[67] Sloane, N. J. A., Hardin, R. H., Duff, T. D. S. and Conway, J. H. *Minimal-energy clusters of hard spheres.* Discrete and Computational Geometry **14**, 237–259 (1995).

[68] Heni, M. and Löwen, H. *Do liquids exhibit local fivefold symmetry at interfaces?* Phys. Rev. E **65**, 9025 (2002).

[69] Gasser, U., Weeks, E. R., Schofield, A., Pusey, P. N. and Weitz, D. A. *Real-Space Imaging of Nucleation and Growth in Colloidal Crystallization.* Science **292**, 258–262 (2001).

[70] Gasser, U., Schofield, A. and Weitz, D. A. *Local order in a supercooled colloidal fluid observed by confocal microscopy.* J. Phys. Condens. Matter **15**, 375–380 (2003).

[71] Volmer, M. and Weber, A. *Keimbildung in übersättigten Gebilden.* Z. Phys. Chem **119**, 277–301 (1926).

[72] Becker, R. and Döring, W. *Kinetische Behandlung der Keimbildung in übersättigten Dämpfen.* Annalen der Physik **24**, 719–752 (1935).

[73] Zeldovich, Y. *Theory of the formation of a new phase cavitation.* J. Exp. Theor. Phys. USSR **12**, 525 (1942).

[74] Turnbull, D. and Fisher, J. C. *Rate of nucleation in condensed systems.* J. Chem. Phys. **17**, 71 (1949).

[75] Herlach, D. M., Egry, I., Baeri, P. and Spaepen, F. *Undercooled Metallic Melts: Properties, Solidification and Metastable Phases.* Proc. NATO ARW 1-2 (1993).

[76] Würth, M., Schwarz, J., Culis, F., Leiderer, P. and Palberg, T. *Growth kinetics of body centered cubic colloidal crystals.* Phys. Rev. E **52**, 6415–6423 (1995).

[77] Wilson, H. A. *On the velocity of solidification and viscosity of supercooled liquids*. Philos. Mag. **50**, 238–250 (1900).

[78] Frenkel, J. *Note on a relation between the speed of crystallization and viscosity*. Phys. Z. Sowjetunion **1**, 498–500 (1932).

[79] Kelton, K. F. *Crystal nucleation in liquids and glasses*. Solid State Physics **45**, 75 (1991).

[80] Williams, R., Crandall, R. S. and Wojtowicz, P. J. *Melting of Crystalline Suspensions of Polystyrene Spheres*. Phys. Rev. Lett. **37**, 348–351 (1976).

[81] Mutaftschiev, B. *Nucleation theory*. Handbook of crystal growth **1**, 189–247 (1993).

[82] Harland, J. L., Henderson, S. I., Underwood, S. M. and van Megen, W. *Observation of Accelerated Nucleation in Dense Colloidal Fluids of Hard Sphere Particles*. Phys. Rev. Lett. **75**, 3572–3575 (1995).

[83] Schöpe, H. J., Bryant, G. and van Megen, W. *Two-Step Crystallization Kinetics in Colloidal Hard-Sphere Systems*. Phys. Rev. Lett. **96**, 175701 (2006).

[84] Collins, F. *Time Lag in Spontaneous Nucleation Due to Non-Steady State Effects*. Z. f. Elektrochem. **59**, 404–407 (1955).

[85] Kashchiev, D. *Solution of the non-steady state problem in nucleation kinetics*. Surface Science **14**, 209 – 220 (1969).

[86] Kelton, K. F. *Transient nucleation in glasses*. Materials Science and Engineering B **32**, 145–151 (1995).

[87] Löwen, H. and Szamel, G. *Long-time self-diffusion coefficient in colloidal suspensions: theory versus simulation*. J. Phys. Condens. Matter **5**, 2295–2306 (1993).

[88] Palberg, T. *Crystallization kinetics of repulsive colloidal spheres*. J. Phys. Condens. Matter **11**, 323–360 (1999).

[89] Stipp, A. *Untersuchungen zur Verfestigungskinetik in Suspensionen kolloidaler Partikel*. Ph.D. thesis, Johannes Gutenberg-Universität Mainz (2005).

[90] Broughton, J. Q., Gilmer, G. H. and Jackson, K. A. *Crystallization rates of a Lennard-Jones liquid*. Phys. Rev. Lett. **49**, 1496–1500 (1982).

[91] Aastuen, D. J. W., Clark, N. A., Cotter, L. K. and Ackerson, B. J. *Nucleation and Growth of Colloidal Crystals*. Phys. Rev. Lett. **57**, 2772 (1986).

[92] Wigner, E. *On the interaction of electrons in metals*. Physical Review **46**, 1002–1011 (1934).

[93] Smith, K. M. and Williams, R. C. *A crystallizable insect virus*. Nature **179**, 119 (1957).

[94] Landau, L. D. and Lifshitz, E. M. *Theory of elasticity* (Elsevier, 1959).

[95] Joanny, F. *Acoustic shear waves in colloidal crystals*. J. Colloid Interface Sci. **71**, 622–623 (1979).

[96] Dubois-Violette, E., Pieranski, P., Rothen, F. and Strzelecki, L. *Shear waves in colloidal crystals: I. Determination of the elastic modulus*. Journal de Physique **41**, 369–376 (1980).

[97] Voigt, W. *Über die Beziehung zwischen den beiden Elastizitätskonstanten isotroper Körper*. Annalen der Physik **274**, 573 (1889).

[98] Johnson, R. *Relationship between two-body interatomic potentials in a lattice model and elastic constants*. Phys. Rev. B **6**, 2094–2100 (1972).

[99] Flügge, S. *Handbuch der Physik Bd. VII, Teil 1 Kristallphysik 1* (Spinger-Verlag Berlin, Göttingen, Heidelberg, 1955).

[100] Raumann, G. *The Anisotropy of the Shear Moduli of Drawn Polyethylene*. Proceedings of the Physical Society **79**, 1221–1233 (1962).

[101] Reuss, A. *Berechnung der Fliessgrenze von Mischkristallen auf Grund der Plastizitätsbedingung fur Einkristalle*. Zeitschrift für Angewandte Mathematik und Mechanik **9** (1929).

[102] Hill, R. *The Elastic Behaviour of a Crystalline Aggregate*. Proceedings of the Physical Society. Section A **65**, 349–354 (1952).

[103] Chaikin, P. M., di Meglio, J. M., Dozier, W. D. and Lindsay, H. M. *Physics of Complex and Supermolecular Fluids* (Wiley New York, 1987).

[104] Schöpe, H., Decker, T. and Palberg, T. *Response of the elastic properties of colloidal crystals to phase transitions and morphological changes*. J. Chem. Phys. **109**, 10068 (1998).

[105] Pieranski, P. *Colloidal crystals*. Contemporary Physics **24**, 25–73 (1983).

[106] Allahyarov, E. A., Podloubny, L. I., Schram, P. P. J. M. and Trigger, S. A. *Damping of longitudinal waves in colloidal crystals of finite size*. Phys. Rev. E **55**, 592–597 (1997).

[107] Berckhemer, H. *Einführung in die Geophysik* (Wissenschaftliche Buchgesellschaft, Darmstadt, 1990).

[108] Ohtsuki, T., Mitaku, S. and Okano, K. *Studies of Ordered Monodisperse Latexes. II. Theory of Mechanical Properties*. Japanese J. Appl. Phys. **17**, 627–635 (1978).

[109] Joanicot, M., Jorand, M., Pieranski, P. and Rothen, F. *Shear waves in colloidal crystals: II. Effects of finite height in cylindrical samples*. Journal de Physique Paris **45**, 1413–1421 (1984).

[110] Phan, S. E. et al. *Linear viscoelasticity of hard sphere colloidal crystals from resonance detected with dynamic light scattering*. Phys. Rev. E **60**, 1988–1998 (1999).

[111] Schöpe, H. J. *Physikalische Eigenschaften kolloidaler Festkörper*. Ph.D. thesis, Johannes Gutenberg-Universität Mainz (2000).

[112] Schöpe, H. J. and Palberg, T. *A Multipurpose Instrument To Measure the Vitreous Properties of Charged Colloidal Solids*. J. Coll. Interf. Sci. **234**, 149–161 (2001).

[113] Palberg, T. et al. *Continuous deionization of latex suspensions*. J. Phys. Chem. **96**, 8180–8183 (1992).

[114] Palberg, T., Schöpe, H. J., Wette, P., Klassen, I. and Holland-Moritz, D. *Solidification experiments in single-component and binary colloidal melts*. Phase Transformations in Multicomponent Melts, ed. D. M. Herlach, Wiley-VCH, Weinheim (2008).

[115] Schöpe, H. J., Decker, T. and Palberg, T. *Response of the elastic properties of colloidal crystals to phase transitions and morphological changes*. J. Chem. Phys. **109**, 10068–10074 (1998).

[116] Latapie, A. and Farkas, D. *Effect of grain size on the elastic properties of nanocrystalline α-iron*. Scripta Materialia **48**, 611–615 (2003).

[117] http://sales.hamamatsu.com.

[118] Scherrer, P. *Bestimmung der Grösse und der inneren Struktur von Kolloidteilchen mittels Röntgenstrahlen.* Nachr. Ges. Wiss. Göttingen **26**, 98–100 (1918).

[119] von Laue, M. *Lorentz-Faktor und Intensitätsverteilung in Debye-Scherrer-Ringen.* Zeitschrift für Kristallographie **64**, 115 (1926).

[120] von Laue, M. *Die äussere Form der Kristalle in ihrem Einfluss auf die Interferenzerscheinungen an Raumgittern.* Ann. Physik **26**, 55 (1936).

[121] Patterson, A. L. *The Diffraction of X-Rays by Small Crystalline Particles.* Phys. Rev. **56**, 972–977 (1939).

[122] Birks, L. S. and Friedman, H. *Particle size determination from X-ray line broadening.* J. Appl. Phys **17**, 687 (1946).

[123] Pecora, R. and Berne, B. J. *Dynamic light scattering* (Plenum Press New York, 1985).

[124] Rička, J. *Dynamic light scattering with single-mode and multimode receivers.* Applied Optics **32**, 2860–2875 (1993).

[125] Hendrix, M. and Leipertz, A. *Photonenkorrelationsspektroskopie.* 3 (WILEY-VCH, Weinheim, 1984).

[126] Palberg, T. et al. *Determination of the shear modulus of colloidal solids with high accuracy.* J. Phys. III France **4**, 457–471 (1994).

[127] Aastuen, D. J. W., Clark, N. A., Swindal, J. C. and Muzny, C. D. *Determination of the colloidal crystal nucleation rate density.* Phase Transitions **21**, 139–155 (1990).

[128] Monovoukas, Y., Fuller, G. G. and Gast, A. P. *Optical anisotropy in colloidal crystals.* J. Chem. Phys. **93**, 8294 (1990).

[129] Monovoukas, Y. and Gast, A. P. *A study of colloidal crystal morphology and orientation via polarizing microscopy.* Langmuir **7**, 460–468 (1991).

[130] Pan, G., Sood, A. K. and Asher, S. A. *Polarization dependence of crystalline colloidal array diffraction.* J. Appl. Phys **84**, 83 (1998).

[131] Roth, S. V. et al. *Small-angle options of the upgraded ultrasmall-angle x-ray scattering beamline BW4 at HASYLAB.* Rev. Sci. Instruments **77**, 085106 (2006).

[132] *http://hasylab.desy.de/facilities/doris_iii/beamlines/bw4/index_eng.html.*

[133] *http://www.marresearch.com.*

[134] *http://pilatus.web.psi.ch/pilatus.htm.*

[135] *http://www.esrf.eu/computing/scientific/FIT2D/.*

[136] Stribeck, N. *X-ray scattering of soft matter* (Springer Heidelberg, New York, 2007).

[137] Wette, P. et al. *Competition between heterogeneous and homogeneous nucleation near a flat wall.* J. Phys. Condens. Matter (accepted, 2009) .

[138] Anderson, V. J. and Lekkerkerker, H. N. W. *Insights into phase transition kinetics from colloid science.* Nature **416**, 811–815 (2002).

[139] Sood, A. K. *Structural ordering in colloidal suspensions.* Solid State Physics **45**, 1–179 (1991).

[140] Konishi, T. et al. *Structural study of silica particle dispersions by ultra-small-angle x-ray scattering*. Phys. Rev. B **51**, 3914–3917 (1995).

[141] Konishi, T. and Ise, N. *Ultra-Small-Angle X-ray Scattering Profile of Colloidal Silica Crystal of 4-fold Symmetry*. J. American Chemical Soc. **117**, 8422–8424 (1995).

[142] Konishi, T. and Ise, N. *Single crystal of colloidal silica particles in a dilute aqueous dispersion as studied by a two-dimensional ultrasmall-angle x-ray scattering*. Phys. Rev. B **57**, 2655–2658 (1998).

[143] Stöber, W., Fink, A. and Bohn, E. *A novel method for synthesis of silica nanoparticles*. J. Coll. Interface Sci. **26**, 62–68 (1968).

[144] Yamanaka, J., Hayashi, Y., Ise, N. and Yamaguchi, T. *Control of the surface charge density of colloidal silica by sodium hydroxide in salt-free and low-salt dispersions*. Phys. Rev. E **55**, 3028–3036 (1997).

[145] Yamanaka, J., Koga, T., Ise, N. and Hashimoto, T. *Control of crystallization of ionic silica particles in aqueous dispersions by sodium hydroxide*. Phys. Rev. E **53**, 4314–4317 (1996).

[146] Gisler, T. et al. *Understanding colloidal charge renormalization from surface chemistry: Experiment and theory*. J. Chem. Phys. **101**, 9924–9936 (1994).

[147] Yamanaka, J., Yoshida, H., Koga, T., Ise, N. and Hashimoto, T. *Reentrant Solid-Liquid Transition in Ionic Colloidal Dispersions by Varying Particle Charge Density*. Phys. Rev. Lett. **80**, 5806–5809 (1998).

[148] Behrens, S. H. and Grier, D. G. *The charge of glass and silica surfaces*. J. Chem. Phys. **115**, 6716 (2001).

[149] Palberg, T. et al. *Shear Modulus Titration in Crystalline Colloidal Suspensions*. J. Coll. Interf. Sci. **169**, 85–89 (1995).

[150] Rojas, L. F., Urban, C., Schurtenberger, P., Gisler, T. and von Grünberg, H. H. *Reappearance of structure in colloidal suspensions*. Europhysics Letters **60**, 802–808 (2002).

[151] Hessinger, D., Evers, M. and Palberg, T. *Independent ion migration in suspensions of strongly interacting charged colloidal spheres*. Phys. Rev. E **61**, 5493–5506 (2000).

[152] Wette, P., Schöpe, H. J. and Palberg, T. *Comparison of colloidal effective charges from different experiments*. J. Chem. Phys. **116**, 10981–10988 (2002).

[153] Cotter, L. K. and Clark, N. A. *Density fluctuation dynamics in a screened Coulomb colloid: Comparison of the liquid and bcc crystal phases*. J. Chem. Phys. **86**, 6616–6621 (1987).

[154] Simonet, V. et al. *Local order and magnetism in liquid Al-Pd-Mn alloys*. Phys. Rev. B **58**, 6273–6286 (1998).

[155] Simonet, V., Hippert, F., Audier, M. and Bellissent, R. *Local order in liquids forming quasicrystals and approximant phases*. Phys. Rev. B **65**, 024203 (2001).

[156] Turnbull, D. *Kinetics of Solidification of Supercooled Liquid Mercury Droplets*. J. Chem. Phys. **20**, 411–424 (1952).

[157] Holland-Moritz, D. et al. *Short-range order in undercooled Co melts*. Journal of Non-Crystalline Solids **312**, 47–51 (2002).

[158] Schenk, T. et al. *Temperature dependence of the chemical short-range order in undercooled and stable Al-Fe-Co liquids*. Europhys. Lett. **65**, 34–40 (2004).

[159] Weeks, J. D., Chandler, D. and Andersen, H. C. *Role of Repulsive Forces in Determining the Equilibrium Structure of Simple Liquids.* J. Chem. Phys. **54**, 5237 (2003).

[160] Herlach, D. M. et al. *Undercoolability of pure Co and Co-based alloys.* Journal of non-crystalline solids **250**, 271–276 (1999).

[161] Harland, J. L. and van Megen, W. *Crystallization kinetics of suspensions of hard colloidal spheres.* Phys. Rev. E **55**, 3054–3067 (1997).

[162] Döbrich, K. M., Rau, C. and Krill, C. E. *Quantitative characterization of the three-dimensional microstructure of polycrystalline Al-Sn using X-ray microtomography.* Metallurgical and Materials Transactions A **35**, 1953–1961 (2004).

[163] Van Megen, W. *Crystallisation and the glass transition in suspensions of hard colloidal spheres.* Transport Theory and Statistical Physics **24**, 1017–1051 (1995).

[164] Ackerson, B. and Schätzel, K. *Classical growth of hard-sphere colloidal crystals.* Phys. Rev. E **52**, 6448 (1995).

[165] Sinn, C., Heymann, A., Stipp, A. and Palberg, T. *Solidification kinetics of hard-sphere colloidal suspensions.* Progress in Colloid & Polymer Science **118**, 266–275 (2001).

[166] Okubo, T., Okada, S. and Tsuchida, A. *Kinetic study on the colloidal crystallization of silica spheres in the highly diluted and exhaustively deionized suspensions as studied by light-scattering and reflection spectroscopy.* J. Coll. Interface Sci. **189**, 337–347 (1997).

[167] Schöpe, H. and Palberg, T. *A study on the homogeneous nucleation kinetics of model charged sphere suspensions.* J. Phys. Condens. Matter **14**, 11573–11587 (2002).

[168] Auer, S. and Frenkel, D. *Prediction of absolute crystal- nucleation rate in hard- sphere colloids.* Nature **409** (2001).

[169] Marr, D. W. and Gast, A. P. *On the solid–fluid interface of adhesive spheres.* J. Chem. Phys. **99**, 2024–2031 (1993).

[170] Marr, D. W. and Gast, A. P. *Interfacial free energy between hard-sphere solids and fluids.* Langmuir **10**, 1348–1350 (1994).

[171] Kelton, K. F., Greer, A. L., Herlach, D. M. and Holland-Moritz, D. *The Influence of Order on the Nucleation Barrier.* MRS Bulletin Research Society **29**, 940–944 (2004).

[172] Bitzer, F., Palberg, T., Löwen, H., Simon, R. and Leiderer, P. *Dynamical test of interaction potentials for colloidal suspensions.* Phys. Rev. E **50**, 2821–2826 (1994).

[173] Blees, M. H., Geurts, J. M. and Leyte, J. C. *Self-diffusion of charged polybutadiene latex particles in water measured by pulsed field gradient NMR.* Langmuir **12**, 1947–1957 (1996).

[174] Van Blaaderen, A., Peetermans, J., Maret, G. and Dhont, J. *Long-time self-diffusion of spherical colloidal particles measured with fluorescence recovery after photobleaching.* J. Chem. Phys. **96**, 4591 (1992).

[175] Nash, G. and Glicksman, M. *A general method for determining solid-liquid interfacial free energies.* Philosophical Magazine **24**, 577–592 (1971).

[176] Glicksman, M. E. and Vold, C. L. *Determination of absolute solid-liquid interfacial free energies in metals.* Acta Metallurgica **17**, 1–11 (1969).

[177] Turnbull, D. and Cech, R. E. *Microscopic observation of the solidification of small metal droplets.* J. Appl. Phys. **21**, 804 (1950).

[178] Hoyt, J. et al. *Crystal-melt interfaces and solidification morphologies in metals and alloys.* MRS bulletin **29**, 935–939 (2004).

[179] Spaepen, F. *Structural Model for the Solid/Liquid Interface in Monatomic Systems.* Acta Metallurgica **23**, 729–743 (1975).

[180] Spaepen, F. and Meyer, R. B. *The surface tension in a structural model for the solid-liquid interface.* Scripta Metallurgica **10**, 257–263 (1976).

[181] Mullis, A. M., Dragnevski, K. I. and Cochrane, R. F. *The transition from the dendritic to the seaweed growth morphology during the solidification of deeply undercooled metallic melts.* Materials Science and Engineering A **375-377**, 157–162 (2004).

[182] Willnecker, R., Herlach, D. M. and Feuerbacher, B. *Nucleation in bulk undercooled Ni-base alloys.* Materials Science and Engineering A **98**, 85 (1988).

[183] Cacciuto, A., Auer, S. and Frenkel, D. *Solid–liquid interfacial free energy of small colloidal hard-sphere crystals.* J. Chem. Phys. **119**, 7467–7470 (2003).

[184] Gránásy, L., Börzsönyi, T. and Pusztai, T. *Nucleation and Bulk Crystallization in Binary Phase Field Theory.* Phys. Rev. Lett. **88**, 206105 (2002).

[185] Hoyt, J. J., Asta, M. and Karma, A. *Method for Computing the Anisotropy of the Solid-Liquid Interfacial Free Energy.* Phys. Rev. Lett. **86**, 5530–5533 (2001).

[186] Li, D., Eckler, K. and Herlach, D. M. *Development of grain structures in highly undercooled germanium and copper.* Journal of Crystal Growth **160**, 59–65 (1996).

[187] McMullen, W. E. and Oxtoby, D. W. *A theoretical study of the hard sphere fluid–solid interface.* J. Chem. Phys. **88**, 1967–1975 (1988).

[188] Broughton, J. Q. and Gilmer, G. H. *Molecular dynamics of the crystal–fluid interface. V. Structure and dynamics of crystal–melt systems.* J. Chem. Phys. **84**, 5749–5758 (1986).

[189] Curtin, W. A. *Density-functional theory of crystal-melt interfaces.* Phys. Rev. B **39**, 6775–6791 (1989).

[190] Ohnesorge, R., Loewen, H. and Wagner, H. *Density functional theory of crystal-fluid interfaces and surface melting.* Phys. Rev. E **50**, 4801–4809 (1994).

[191] Davidchack, R. L. and Laird, B. B. *Direct Calculation of the Hard-Sphere Crystal /Melt Interfacial Free Energy.* Phys. Rev. Lett. **85**, 4751–4754 (2000).

[192] Sun, D. Y., Asta, M. and Hoyt, J. J. *Crystal-melt interfacial free energies and mobilities in fcc and bcc Fe.* Phys. Rev. B **69**, 174103 (2004).

[193] Oxtoby, D. W. *Homogeneous nucleation: theory and experiment.* J. Phys. Condens. Matter **4**, 7627 (1992).

Publications within the thesis

2009

- Klassen, I., Wette, P., Holland-Moritz, D., Palberg, T. and Herlach, D. M. *Short-range order of tunable colloidal suspensions: from soft to hard sphere like behavior*, submitted

- Wette, P., Klassen, I., Lorenz, N., Holland-Moritz, D., Palberg, T. and Herlach, D. M. *Complete description of re-entrant phase behaviour in a charge variable colloidal model system*, submitted

- Lorenz, N., Schöpe, H. J., Reiber, H., Palberg, T., Wette, P., Klassen, I., Holland-Moritz, D., Herlach, D. M., and Okubo, T. *Phase behaviour of deionized binary mixtures of charged colloidal spheres*, J. Phys. Condens. Matter **21**, 464116 (2009)

- Wette, P., Engelbrecht, A., Salh, R., Klassen, I., Menke, D., Herlach, D. H., Roth, S. V., and Schöpe, H. J., *Competition between heterogeneous and homogeneous nucleation near a flat wall*, J. Phys. Condens. Matter **21**, 464115 (2009)

- Wette, P., Klassen, I., Holland-Moritz, D., Palberg, T., Roth, S. V. and Herlach, D. M. *Colloids as model systems for liquid undercooled metals*, Phys. Rev. E. **79**, 010501(R) (2009)

2008

- Palberg, T., Lorenz, N., Schöpe, H. J., Wette, P., Klassen, I., Holland-Moritz, D. and Herlach, D. M. *Solidification experiments in single component and binary colloidal melts*. in: Phase Transformations in Multicomponent Melts, ed. by D. M. Herlach (Wiley-VCH, Weinheim 2008)

2007

- Wette, P., Klassen, I., Lorenz, N., Klein, S., Holland-Moritz, D., Palberg, T. and Herlach, D. M. *Determination of the short-range order in undercooled colloidal melts.*, HASYLAB Annual Report, 1339 (2007)

Danksagung

Mein besonderer Dank gilt Prof. Dr. D. M. Herlach für die interessante Aufgabenstellung und die Betreuung dieser Arbeit. Seine lehrreichen und motivierenden Ratschläge sowie seine Förderung machten es mir möglich wertvolle Erfahrungen in den letzten drei Jahren zu sammeln.

Prof. Dr. A. Meyer danke ich für die Möglichkeit zur Durchführung dieser Arbeit am Institut für Materialphysik im Weltraum (DLR, Köln).

Prof. Dr. Dr. h. c. H. Zabel danke ich für die Bereitschaft das Koreferat zu überneh- men.

Ein ganz besonderer Dank gilt Dr. P. Wette, der die Brücke zwischen der Kolloid- und Metallphysik an diesem Institut geschlagen hatte. Ihm ist es zu verdanken, dass sich ein neues Forschungsgebiet innerhalb kürzester Zeit etablierte und in guter Zusammenarbeit mit Prof. Dr. T. Palberg und Priv.-Doz. Dr. H. J. Schöpe von der Universität in Mainz vielversprechende gemeinsame Projekte ins Leben gerufen werden konnten. Sein unermüdlicher Einsatz für die Wissenschaft und Forschung war stets bewundernswert und wirkte sehr motivierend.

Priv.-Doz. Dr. D. Holland-Moritz möchte ich für die Unterstützung dieser Arbeit in Form von Simulationsrechnungen der Strukturfaktoren und die ständige Bereitschaft zu Diskussionen danken.

Ausdrücklich bedanken möchte mich an dieser Stelle auch bei LAss. H. Reiber und D. Menke für die exzellente Probenpräparation.

Dem BW4 team am HASYLAB, insbesondere Dr. S. V. Roth und Dr. A. Timmann, danke ich für die Unterstützung der USAXS Messungen. Begleitet wurden die Messkampagnen von N. Lorenz, S. Klein und D. Menke, denen ich ebenfalls für den unermüdlichen Einsatz samt Nachtschichtarbeit an der BW4 danken möchte.

Für die finanzielle Unterstützung bedanke ich mich bei der Deutschen Forschungsgemeinschaft, die diese Arbeit im Rahmen der Schwerpunktprogramme SPP1120 (HE1601/16) und SPP1296 (HE1601/24) gefördert hat.

Und ganz zum Schluss möchte ich mich bei der gesamten Arbeitsgruppe samt ehmemaligen Mitarbeitern für das angenehme Arbeitsklima bedanken, das viel dazu beigetragen hat, diese Arbeit immer in guter Erinnerung zu behalten.

Die VDM Verlagsservicegesellschaft sucht für wissenschaftliche Verlage abgeschlossene und herausragende

Dissertationen, Habilitationen, Diplomarbeiten, Master Theses, Magisterarbeiten usw.

für die kostenlose Publikation als Fachbuch.

Sie verfügen über eine Arbeit, die hohen inhaltlichen und formalen Ansprüchen genügt, und haben Interesse an einer honorarvergüteten Publikation?

Dann senden Sie bitte erste Informationen über sich und Ihre Arbeit per Email an *info@vdm-vsg.de*.

Sie erhalten kurzfristig unser Feedback!

VDM Verlagsservicegesellschaft mbH
Dudweiler Landstr. 99   Telefon +49 681 3720 174
D - 66123 Saarbrücken   Fax   +49 681 3720 1749
www.vdm-vsg.de

Die VDM Verlagsservicegesellschaft mbH vertritt

Printed by Books on Demand GmbH, Norderstedt / Germany